No Risk Ranching

Custom Grazing on Leased Land

No Risk Ranching

Custom Grazing on Leased Land

by

Greg Judy

A division of Mississippi Valley Publishing Corp.
Ridgeland, Mississippi

First printing August 2002
Second printing March 2003
Third printing February 2006
Fourth printing June 2008
Fifth printing December 2009
Sixth printing May 2011
Seventh printing March 2013
Eighth printing November 2015
Ninth printing May 2017
Tenth printing November 2018
Eleventh printing August 2019

Library of Congress Cataloging-in-Publication Data

Judy, Greg, 1959-
No risk ranching : custom grazing on leased land / by Greg Judy
p. cm.
Includes bibliographical references (p.).
ISBN 0-9632460-8-9
1. Grazing--Management. 2. Grazing--Economic aspects. 3. Small business. I. Title.

SF85 .J84 2002
636.08'45--dc21

2002072381

Cover Design by Steve Erickson, Ridgeland, MS

Manufactured in the United States of America

This book is printed on recycled paper.

I am dedicating this book to Steve Baima. He is the person who pointed me in the right grazing direction! Steve took the time to explain the concept of Management-intensive Grazing to me. One of the best pointers he gave me was to subscribe to the *Stockman Grass Farmer*.

Acknowledgments

I would like to thank the following people for their wisdom, help and valuable information that they shared with me over the years.

Jan Judy (My wonderful wife who has helped me grow our grazing company enormously over the past several years). Allan and Virginia Judy (my parents), Dan Judy (my uncle), Chris and Angie Roberts (nephew and niece), Allan Nation, Jim Gerrish, Fred Martz, Maurice Davis, Rob Kellenbach, Craig Roberts, Casey Olsen, Steve and Cindy McCawley, Mike Curry, Marshall Colley, Bill Rinehart, Cheryl Livengood, Cindy Bowne, Jack and Dulane Wohlman, Frank Bowen, and Barb Turnball.

Table of Contents

Foreword

Custom Grazing with Leased Land = $$$$$

My name is Greg Judy. I am 42 years old and live in Rucker, Missouri, almost in the dead center of the state. I was born on a dairy farm in northern Minnesota. Parents included, there were nine of us.

Our family moved to Missouri in 1966, mainly to escape the cold long winters. My Dad sold the dairy cows shortly thereafter and went to work in town. He kept one cow for the family milk needs. I used to watch in amazement how my dad blasted the milk streams into the bucket with his powerful hands and with such great rhythm. When I reached seven years of age I asked him one night if I could try milking the cow. From that point on, I had the job until I moved away from home after high school!

I remember the freezing mornings, how the back of your fingers and hands would freeze as you tried to get done as fast as possible. The cow was always kicking at biting flies, and runny manure-laden tails would smack you in the face while you tried to milk in the hot summer months. After milking I let the milk cow's calf have half of one hind teat. You had to wrestle the calf away from the cow before turning the cow out to grass. As the calf grew it got to be quite a chore. You grabbed him by an ear and the tail and with a lot of momentum because the calf always out-weighed you. He was never ready to leave.

After the milking I would strain the milk through a fine cloth to remove any debris. I remember my grade school principal telling me that it was unsafe to drink the raw milk; we were supposedly at risk of contracting all kinds of diseases! I asked him what he drank while he was growing up? I never heard anymore from him about unsafe milk. It was an extremely satisfying feeling to head to the house with a bucket full of fresh milk for the family. The cow was always ready to go back to grazing. The calf was full. The barn cats had a little milk. And I usually poured a little out for the meat hogs on the way to the house. Life was good.

As a youngster there were times when I was milking the cow that I would have rather have been doing about anything else. But guess what? You couldn't pay me enough to give up that experience now. It taught me responsibility, a work ethic, and a daily job that had to be done twice a day by me or it would not get done. I know this is not possible, but it would be fantastic if every kid had some type of chores with an animal every day.

At age 17 I went to work as a carpenter building houses for a contractor. At age 20 I worked for an electric utility supplier, working in a factory. I worked in the factory for ten years, then moved up to engineering and product development for another ten years. I do not have a degree, only a lot of first-hand practical experience.

Between 1993 and 1995, I bought my Uncle and Dad's farms totaling 205 acres. This consisted of two 60-acre plots and one 75 acre plot. I paid $350-$400 per acre, which I thought was a good price at the time. I was able to scrape enough money together to pay 10% down on all plots. They all join each other with my sister and brother-in-law's 40 acres thrown in the middle, which I also pasture. The farm had been severely neglected, with mainly large cedars, blackberry thickets, wild plum thickets, sumac, and dewberry vines. It is all rolling hills, with a heavy clay base. There is one to three inches of topsoil on most of it.

The land had been heavily cropped earlier in the century, and basically all the topsoil washed down into the Missouri River. I talked to an old timer who had plowed and cultivated my farm in the 1920s and 1930s. He commented that he used a team of mules and plowed downhill, because it was easier on the mules. He told me that he raised a fair crop of corn on it for about two years and then the old place just played out, and he couldn't raise anything but nubbins! I just can't imagine why it would have done that.

I figured that as long as my job in town held out I could make the farm payments without starving to death. Plus I figured with all this land, in no time I would have it packed with cattle and I would be getting rich! I didn't know beans about grasses or the importance of legumes. I was about to learn a big lesson in the life of the evil interest payments and cattle cycle.

I cleared all the ridge tops with a chainsaw and painted all the stumps. I bought 16 bred cows over a two-year period, (at the cattle cycle peak!). By the middle or end of July I was out of grass, strictly because I was grazing the whole farm at once and had poor fertility to go along with it.

A neighbor of mine introduced me to a fellow who had his farm set up in multiple paddocks. I will never forget it. We were in the middle of a drought when we walked his farm. He had 40-50 cow-calf pairs on 160 acres and the most gorgeous grass/legume mix you ever saw. My 200 acre farm looked like a pool table with 16 cow-calf pairs! You talk about an eye opener. At first I thought, "Well he has got better ground than I do. My farm will never be able to look like this."

This fellow sat down and explained the concept of MiG

and suggested that I go to the Linneus Grazing School, which is a University of Missouri research farm that teaches graziers about forages, soils, livestock, etc. He also suggested that I start reading a magazine called the *Stockman Grass Farmer*. He lent me an old copy. I devoured every article and immediately subscribed to it. At that time I was scraping the bottom of the barrel to make the interest payment on the farm and pay a little toward the principle. I could see that if I did not change something I would be in debt the rest of my life trying to pay off this farm.

I remember driving up through the research farm on the way to the three-day school and drooling over the gorgeous paddocks of grass. I slept in my camper shell every night because I did not have the money for a hotel room, and barely had enough to pay for the grazing school. That first night I got trapped by floods going into the research station. I drove through water until it started coming up under the door. I found a high spot and camped the night right there. It rained buckets every day but that did not dampen my enthusiasm one bit.

That was the best three days I ever spent learning about how to grow forages, build fences, manage herd health, you name it, and they taught it. They had a whole panel of professors, each with his own specialty. I was in heaven, and could not wait to get home and start stringing wire for paddocks and putting some of this new thinking to work.

I signed up for a cost/share program through our local county Soil Conservation Service office to do a complete pasture re-establishment program. At the same time this Allan Nation fellow kept warning everyone of the coming cattle cycle crash in his editorials in the *Stockman Grass Farmer*. I had just kept back a nice group of raised heifers and was getting ready to increase my herd.

Cow prices had already dropped some, but I sold everything I had, bought back all steers in February of 1996. It was one of the hardest things I ever had to do. I loved my cows, even though they were starving me to death! Shortly afterwards cow prices plummeted, which helped ease the pain of selling them.

I actually made enough money that summer grazing those steers to make a farm payment. That same summer disaster hit. I discovered my wife was a manic depressant. She filed for divorce. Luckily there were no kids involved. Now, I had two drains on my finances, the divorce and the farm/house payment. All cash was quickly sucked up by the divorce and land payments.

I had no money and no cattle to generate any money. I was at the bottom of the barrel and figured my farm was history also. My situation looked hopeless. There was no end in sight for the divorce proceedings either. It seemed as if the whole world was against me. During that winter the same fellow who introduced me to MiG came up to me and asked me if I would be interested in grazing his yearlings for the upcoming grazing season. I jumped at the offer to graze them. This allowed me to stock the improved paddock portions of the farm. From April 20th to the end of July those calves averaged 2 lbs a day gain, 200 lbs of gain per head in 100 days.

I put an ad in some local papers looking for some livestock to graze on the un-improved portions of the farm also. I was looking for dry cows, but a rodeo owner called me and asked if I would be interested in grazing his bucking horses for him. I wrote out a one year contract and started grazing horses. This was a good learning experience, and it also paid very well. The horses cleaned up all my brushy areas and ate everything except the thorns! I made up six paddocks and rotated 30 horses through them. I never realized that a horse eats all the time, but I somehow managed to keep them in grass and brush.

That allowed me to make the first six-month farm payment. Now if only I had more land to run more stockers on, I thought. Then I read an article by Allan Nation that preached on the theory that your sole purpose should not be to own the land, but to make a living from the land. The whole article hit me like a brick in the forehead. I started driving around the neighborhood eyeballing all the idle pastures.

I now have ten farms including my own totaling 1460 acres. 900 acres is in grazeable grass, the rest is timber and unfenced areas yet to be developed. I went from 40 stockers in 1996 to 1100 custom grazed stockers in 2001. None of this growth would have been possible without custom grazing and leasing idle land. I was dead broke and had no money to expand. There is no stress, because I don't own anything. I am effectively selling my management skills to develop idle land and put economic grass gain on stockers.

One of the main driving forces for me writing this book is to prevent other young potential graziers from making the same mistakes that I made. I was so broke from trying to pay the land interest payment that I could not ever get ahead enough to make any major principle payments. Leasing farms and custom grazing has made it possible for me to pay off my own farm and house in three years. If you believe in something strongly enough and are committed to making it happen, it will happen.

I am convinced that in the USA our pastures are one of our most underutilized natural resources. I am not against land ownership either; I just feel like it is an awful hardship on a new blooming grazing business. I have nothing against owning cattle either. I prefer custom grazing starting out, because it frees up your equity to grow your business. Go out and get your arms around as much of it as you can.

This book will focus on all aspects of custom grazing and leasing land. I believe you will find this book very informative and if you apply the same principles, it will make you a successful, profitable grazier also. To me there is no other more noble vocation then putting gain on young calves with tender grass. It just doesn't get any better than that!

Pasture Leasing & Custom Grazing

Leasing idle land for pasture is the best way for a grazier to build equity and produce a profit year after year. There is absolutely no better way for a young grazier to get started in the art of mastering your grass growing knowledge, grazing skills, managing livestock than to go rent some pasture.

There is so much doom and gloom circulating through present day agriculture about ground water contamination, low commodity prices, adverse weather conditions, pests, government regulations, taxes, tainted beef, the European beef ban, and the list goes on and on.

A refreshing break from all of this is that there are millions of acres available in the USA not being utilized for anything. This is where our opportunity exists. I will discuss in this book, how to find it, secure a long-term lease on it, and develop it for a profitable grazing operation.

If you want to position yourself to make a decent living in grass farming there is absolutely no better risk-free, stress-free way to do so. Depending on how good of a grass manager you become there are some very handsome profits to be made with leased pasture. You must use Management-intensive Grazing (MiG) to maximize the highest potential from your leased pastures. No other form of grazing will turn a pasture around and be as profitable as fast as mob grazing.

Chapter 1

Forming the Correct Personal Attitude

To be able to convince people to let you lease their land, it takes a very honest, positive and optimistic attitude. I like being around optimistic people, they have this energy that seems to glow. Optimism is contagious. You have to open up your shell and get infected by it. Here are some tips for becoming more of an optimist:

1. The next time a negative thought enters your mind, immediately focus your mind on replacing it with a positive thought. Work at this constantly.

2. Learn to love the challenge of treading on uncharted ground. These are the fun times of your business, enjoy them, rather then worrying about, "What if?"

3. When faced with your next negative situation, adopt the attitude, "It is only as negative as I make it, life goes on!"

4. Focus on things that you can control: your own actions, securing leases, water development, paddock divisions, keeping grass vegetative.

5. Don't let your mind worry about the things that you cannot control: the weather, pests, commodity prices, what people think of you.

6. Avoid the pity party trap. People feeling rotten about themselves and life would love to have your company.

7. Never accept defeat. You will make mistakes. Learn from them and go on. The only way you can fail in life is to make the decision, "I'm a failure and I absolutely refuse to try anymore." Failures are part of the learning process, they build character, humility, strength and wisdom.

8. Surround yourself with successful people. Study them and their lifestyle.

Adopt a Can Do Attitude

Life is a series of bumps, then smooth spots, then more bumps. The "Unknown" is what makes life exciting and gives us something to work toward. If everything we tried in life worked, life would be boring. We would soon have the attitude, "Why try it? I know it's going to work."

The really painful things that didn't work are the ones that I remember the best.

When I first started leasing land and custom grazing, I would have people ask me, "How in the world do you expect to make any money when you don't own the land or the cattle?"

Every time someone said this to me, I would use it as an inspiration to concentrate on obtaining as much cheap grazeable grass as I could get my hands on. Most of the people who asked me this question were farmers who had tractors, balers, mowers, stock trailers, new trucks, a new ATV, and maybe a loan against the cow herd.

Their reasoning was, "Heck we haven't had to pay taxes in 30 years, because our farm has not shown a profit and we own all the equipment. You're out of your mind if you think you're going to make money by not owning nothing!"

I did not let their negative comments dampen my optimism one bit. What they didn't realize was that I had minimal overhead, no depreciation, no property taxes or insurance, no risk of death loss, no risk of livestock or land prices dropping, and my capital was freed up to grow the business. I could concentrate solely on finding economical leases, growing tender vegetative grass and lining up stockers to graze.

You have to be firmly convinced that there is no cheaper way to put a pound of gain on an animal than by grazing grass. I knew that I had an unfair advantage because I could put gain on cheaper than the conventional method (grain).

You have to become a self-motivator, and set realistic attainable goals. Your brain is your most powerful tool. Don't waste it and let it lay idle. A human brain is never fully developed. It is like a muscle. The more you use it, the bigger it gets. Don't let your mind get in a rut of constantly doing the same task, the same way every time. Stand back and look at what you are doing and ask yourself, "Is there a more efficient way to do this?" I do this constantly and sometimes I get a green light that turns on, "Why didn't I think of that before?"

Once you've tackled a difficult goal and are successful at it, you're on the way up. This builds tremendous self-confidence. Once you learn the proper ingredients to be successful, just keep repeating the recipe.

Let's Be Friendly Folks

Always look for the best in people instead of their shortcomings. Make one new friend a day and you have left behind a potential enemy. Everybody has faults, nobody is perfect, but by not dwelling on the negative side of people and focusing on the positive side you can make a lot of new friends.

Do your very best to get rid of one enemy a day and pretty soon you won't have any. It is a lot better for you physically and mentally to get along with people than to antagonize them. I constantly wave at people when I drive around the neighborhood. Sometimes you get some strange looks, but pretty soon those people start waving back. If a neighbor asks you to help him, never turn him down. You will need his help some day and this also builds strong friendships and close knit communities.

It makes me feel good to be friendly and have people be friendly back. I've had people riding with me comment, "Well, Greg, who was that?"

I will reply, "I don't know."

Then they will reply, "Well, why did you wave?"

"Just to be friendly," is my reply.

The world would be an awfully friendly place if people got in the habit of waving at each other. What people think of you, I believe, is formed when you have your first contact with them. Rehearse exactly what you want to say before you meet with the landowner.

Be polite. Use no profanity. Wear clean clothes. Be clean-shaven, with a trimmed beard. And preferably, have a clean vehicle. Offer a firm, but not bone-crushing handshake to the landowner while giving him eye-to-eye contact. This is extremely important. It builds trust. There is nothing more disappointing than a limp-fish handshake. A wimpy handshake is a real business turn-off. The landowner is thinking, "This guy has not even got the gumption to give me a decent handshake, how is he going to manage my property?" With most people you meet, impressions are formed from a mere handshake. So practice your handshakes with your wife or a friend!

Your intention is to manage the landowner's land as if it were your own.

Keep eye contact when you are talking to them. I'm sure you have heard people say, "If a man won't look me in the eye when he is talking to me, then I don't trust him." Be careful not to talk too much or you may form the impression that you are a smart-aleck know-it-all type.

An opening statement to a landowner might sound something like this, "I was wondering if you would be interested in leasing your open pasture for grazing?"

The landowner will either reply yes, no or maybe. If they say no, there is probably a good reason for it. Maybe the landowner had a bad experience in the past by leasing it, or they're planning to develop it. They might be selling it, or maybe they feel uncomfortable about having livestock on their land.

Concentrate on being a positive, optimistic person.

If you get a yes or a maybe for an answer, then ask the landowner if you could walk the property with them to examine its grazing potential. Ask them what their concerns are with the property. They may have a beautiful lake smack dab in the middle of the pasture and all they can envision is your cattle taking baths in it every day.

They may have a stand of pine or nut trees that they planted and nurtured.

Ask them how they obtained the land. If you know any history of the land, tell them about it.

Ask what their long-term plans are for their property. Did they buy it as an investment, to hunt on, or possibly to build that retirement home on?

You have to be excited when you find a piece of ground that interests you, then it is contagious and the owner gets excited. Do not immediately spill out a whole list of promises, but be honest about what you can do for the property to increase its value.

When you start talking in detail you want to have everything written down in the order that it will be performed – securing the perimeter fence, building soil fertility, developing water sources, etc. I will go into more detail about the actual written contract later in the book.

Some landowners are very cautious when you first meet them to talk about their land.

Take a good long look at all the shortcomings that the land has and figure cost effective ways to fix them. Some examples would be washed out spillways, weed problems, brush invasion, broken down perimeter fences, neglected wood lots.

Offer to trim all low branches off walnut trees, perform timber stand improvement practices, fence off creeks that are eroding out into the fields, etc. This shows the landowner that you are genuinely concerned about his property. If he has any woods on the property, explain to him that you will fence those areas off to prevent cattle from tromping valuable wildlife cover and marketable timber in the future.

Most landowners are emotionally attached to their property and you need to express to them that you are concerned also. Explain that it is your intention to manage his land like it was your own.

As you go along and get to know the landowner better, some tremendous friendships are built. I personally get a high degree of reward from seeing a satisfied landowner grinning and shaking my hand, commenting on how beautiful their farm looks.

Above all, be honest with the landowner. If you make a mistake, own up to it. He will admire you for your honesty. Manage his ground as if it was yours and your livelihood depended on it. Someday it may.

Go the extra mile to satisfy any concerns the landowner has. Write his concerns down, revisit your notes throughout the lease to make sure you are still addressing his concerns.

Chapter 2
Setting Goals

I began setting goals during my divorce settlement in 1996, so no one else was involved. Up to that point I had never written down anything to work toward. Once I started writing down my goals, good things started to happen, all because I had a plan to follow.

If you have a wife and family, include their thoughts in setting goals. This builds a team concept, everyone is working toward the same goal.

Don't let anybody try and convince you otherwise. Be an optimist about everything in life. Hey you only get one chance at living your life, don't waste it!

If you are doing everything like your neighbors are, you are probably not going to reach your goals.

Avoid negative talking people. They will try and convince you that what you are doing is wrong. When a person tells you that it cannot be done and you go out and do it anyway and are successful at it, it just makes them even madder. I've had people tell me, "Leasing land may work in your area, but it won't work in mine."

I reply, "Well, maybe you could look in another area."

You may need to be like a hound dog and get out there and sniff it out.

I find that it helps to set quarterly goals, then you can see how close you are to reaching them. By setting quarterly goals it makes reaching your end goal a lot easier. It's kind of like climbing a ladder, step by step. By biting off small pieces, the goal doesn't seem so impossible to reach.

Revisit your goals quarterly to see how close you have stayed on course. Don't let anybody tell you that you will not reach your goal. The only person who can prevent you from reaching your goal is yourself.

How I Reached My Goals

My first goal was to pay off my house in one year.

I realized that if I was going to reach this goal, there had to be some outside income from somewhere besides my off farm job.

The one sentence from Allan Nation that kept coming back to me and made so much common sense was, "Your goal should be to make a living from the land, not to own it all." It really ate at me that I had done exactly the opposite of that. I also read Joel Salatin's article in the *Stockman Grass Farmer* about the huge opportunity facing young graziers today, if they were willing to get out and learn MiG and implement it. I can not over emphasize what a strong tool MiG is.

I started leasing land and custom grazing with very little equity involved. I didn't have any money. The first year I got three farms leased very economically because of lack of fencing and the land laying idle. I started custom grazing immediately and generating a monthly income from that. I secured a rodeo horse grazing lease, leased out the hunting rights to my own farm, cut the marketable timber on my farm and watched my spending.

All equity was paid toward the house loan. I met my goal of paying off the house loan in one year.

Next I set a goal to pay off the farm loan in two years. Once I had the goal written down, every spare dollar went toward a land principle payment. I paid off the $42,000 farm loan in 20 months.

Revisit your goals quarterly to see how close you have stayed on course.

Another goal I set is to own at least half of all stockers that I graze each year. I will watch for the cattle cycle bottom. I will leverage my owned farm to buy the stockers. This way I own the cattle, and the banker can not tell me when to sell and buy.

I have found that lending institutions are more than willing to lend you money when cattle prices are at the peak of the cycle. Emotions are running high. Everybody wants a piece of the action. This is absolutely the worst time to borrow money to buy cattle, because all they can do is drop in price.

My first custom stocker grazing goal was to reach 500 stockers per year. In my area on unimproved clay base rolling pastures you can calculate about 0.8 to1 steer per acre for the April through July period. Then you rest pastures 45-60 days and run the fall group, assuming you get some rain for re-growth.

I figured that I needed 300-350 acres of grass to reach my goal of 500 stockers per year. I went out and got busy finding land to lease close to my farm. Once that goal was reached, a new goal was set of grazing 1000 stockers per year.

Now some of my farms had been improved to the point to where I could run 1 to1.5 stockers per acre. I roughly figured that I needed 700-800 acres of grass if I was to reach my goal of 1000 stockers per year. I have reached this goal, so I set a new goal of being able to graze 1500 stockers full time and quit my off farm job by age 45. I am confident that this goal is attainable as long as I stay focused on it. I have found that if you truly believe it will happen, by gosh it will.

Some Realistic Goals to Work Toward

1. Where do you want to be financially in 5 years?
2. Set a goal to make a profit every year.
3. Find one farm that is capable of grazing 80-100 stockers; secure an economical lease on it and develop a MiG system.
4. Set a goal to own at least half of the stockers that you graze.
5. Establish strategic and economical watering sites on the leased farm.
6. Establish legumes as quickly as possible.
7. Develop your grass growing and grazing skills.
8. Never be satisfied that you have learned everything.

When you reach a goal, reward yourself. Take a vacation. Don't be afraid to set goals too high. It's better to set your goals high as this prevents you from selling yourself short. I guarantee you will be further along if you set goals, than if you never set any.

One of the keys to setting goals is to look in the future. Imagine where you would like to be financially in five, ten, fifteen years. Next decide how much income you need to make in order to have a comfortable living from your grazing operation. To a lot of people this would mean the income needs to equal or exceed what they are presently making with their off the farm job.

The only person who can prevent you from reaching your goal is yourself.

This is how I calculate my potential future income:

Let's say that I decide that I need $35,000 per year to

live on (this would be the off farm salary).

With 500 custom grazed stockers I could achieve this goal. Let's look at the arithmetic:

* 500 stockers x 240 days grazing x 1.2 lbs gain/day x .33 cents (custom charge for each lb of gain) = $47,520.00 - $12, 520 (lease, gas, truck repair,etc) = $35,000 profit

Once you have decided on your income and how many stockers you will need to achieve that level of income, next you have to calculate how many acres of grass will be needed to run 500 stockers for 240 days.

Let's assume you need one acre of grass per stocker, which means you would set your goal at finding 500 acres of leased pasture. You can plug in any set of numbers with the above example, but the principle remains the same.

Write Your Goal Down and Keep It Handy

If you are going to be successful in life, it helps to write down a set of goals. Writing it down is very important. If you physically write it down and post it where you can read it, this helps keep you focused on reaching your goal.

I keep my present goal posted on the refrigerator, bathroom mirror, bedroom mirror, and in my wallet. Everywhere I look, there is my goal staring right back at me! You have to firmly believe that putting gain on a young animal with tender succulent grass is absolutely the cheapest gain available. This one single fact was anchored in concrete in my mind. No one could persuade me otherwise. This has really helped in achieving my goals.

Another tool that helps keep you focused on your goals is to carry a pocket notebook pad and pen with you at all times. I personally prefer the Day-Timer pad. It has two pages for each day of the year with labeled spots for expenses, to-be-done-today action list, a diary and work record, space for appointments and scheduled events. In the Day-Timer package you get a new one for each month along with a mini filing case for permanent record keeping.

It is very important to keep accurate written records. I am constantly referring back to my notebook for all kinds of past information. If you write it down it will get done, it's that simple. It is also nice when you are dealing with several different landowners to refer back to your written notes; it prevents confusion. I am constantly going back to my mini filing cabinet where each previous monthly Day-Timer is stored. I can instantly refer to notes and important information that is stored in them. Each day I write down my action to do list, a simple four to five word description per item.

Actions are written down in the order of importance (I like to call it "My Biggest Bang For The Buck"). If one is not completed that day, I move it to the next day's action list. Sometimes you get a neat idea or see something that interests you; write it down so you don't forget it.

Chapter 3

Advantages of Leasing Versus Owning Land

It seems in America that most young prospective ranchers start out with the dream of owning their own spread.

They go to the bank and scrape enough money together to come up with the down payment then spend a lifetime trying to pay for it. The problem with this scheme is that the bank owns you, your cattle, machinery, etc. There is never enough profit left to make a sizable principle payment on your land loan, let alone to have enough to live on. I know first hand because I took this painful route.

Let's talk about that awful word "Stress." You greatly minimize stress or completely eliminate it by leasing land and custom grazing. You are not responsible for any of the payments for owning the livestock or land. I can remember cutting and selling firewood so that I would have enough money to buy groceries, because the farm and cattle loan payment kept me broke. It was taking all the money that I earned in town to simply try and live on the farm. I was not a big spender either. It's not a real good feeling to work all year and end up giving all your hard sweat earnings to the banker. He smiles and says, "You come back and see me next year."

If I was given the opportunity to start all over again, I would never buy land at the beginning of a fledgling grazing business. You are doomed for failure before you can even get your feet on the ground and learn the grazing business. To quote Allan Nation, "Livestock ownership should always come before land ownership." If you buy a farm, usually there is not any extra money left to purchase any livestock after you make the interest payment. Remember the goal is to make a living from the land, not to have it own you for your whole lifetime (with interest payments).

> *Livestock ownership should **always** come before land ownership.*

This is where leasing land comes into the equation. When you lease land, the cost of the lease is 100% tax deductible, (it's a business expense), unlike owning land where you are not allowed to deduct principle payments from your income. There are no property taxes to pay at the end of the year with leased land. There's no insurance on the land. Remember you don't own it.

With leased land you are free from the huge interest payments that kept you broke. There is basically no overhead that keeps your finances constantly drained. You are free from any liability if someone gets hurt on the leased land. You don't have to worry if the price of land is going up or down, because you don't own it. Any acts of nature, tornadoes, floods, etc. will not break you. It may be a mess to clean up, but you are not out any financial loss.

The biggest advantage of leasing is it allows you to build up enough equity to buy your own livestock and learn MiG as you go, so if and when you decide to purchase some land you have a cash-generating herd paid for to help you with your land payments.

An Example of Leasing Versus Owning

My leases range from two to twenty dollars per acre per year. I have one 40 acre solid grass farm leased for $400 per year or $10.00 per acre. This farm would bring $1500/acre on the open market if it was sold. So for a $400 a year lease, I am controlling a $60,000 grass farm.

Now let's assume I decided that I wanted to buy it and pay for it with the livestock that were grazed on it. First I would have to come up with a down payment of $6000, 10% down, then find a bank that will loan me the rest of the money at a decent interest rate. Let's say we are able to secure a $54,000 loan at 8% interest and we also come up with the down payment of $6000.

We will have an annual interest payment of $4320 the first year. In this example we will custom graze and charge 32 cents per pound of gain. The grass farm will stock one steer per acre, which equals 40 stockers. Let's assume the best grazing season ever hits that first year and your stockers gain 300 lbs. At $96 gross per stocker x 40 head = $3840 gross custom grass gain.

We have to subtract grazing expenses, gas, salt, truck repair, etc. from our gross gain. We will give those a total of $400. Take $3840 - $400 (grazing expenses) = $3440. Now we have to make the land interest payment of $4320. We are $880 short. We did not make enough to make the interest payment of $4320, let alone any principle or living expenses.

We still have those land taxes and insurance to pay at the end of the year.

Now let's go through the same exercise pretending that we leased the land for what I have it leased for. The gross custom grass gain would be the same $3840. Now we subtract our lease of $400 per year. This leaves us with $3440 - $400 for grazing expenses = $3040 profit. Which scenario do you want to try and make a living from? Owning the land or leasing it?

Leasing Allows Quick Economical Growth

Leasing land allowed me to grow my grazing business from my owned land of 205 acres, to 1300 acres in just three years. Of this 1100 leased acres, 800 was grass, the rest in woods and brush. There is absolutely no other way for a grazier to get control of this much ground economically in such a short period of time. To buy the additional 1100 acres would have cost 1.65 million dollars (at $1500 per acre). It would be difficult to graze enough stockers to ever pay for it. If I had tried to expand my grazing entity by buying the additional 1100 acres that I leased, here is what the numbers would look like:

1. 10% down payment = $165,000
2. Secure loan at 8% interest = 1st year's interest is $118,800
3. Land owner liability insurance and taxes per year = $5000
4. Farm has fair grass, we will assume .875 steer per acre (very liberal for first year!).
5. Our custom grazing fee will be $0.32 per pound of gain.
6. Custom graze 700 stockers and put 300 lbs gain on each animal.
7. Grazing expenses $5.00 per stocker = $3500.
8. $96 x 700 stockers = $67,200 gross income.
9. Take $67,200 - $8500 insurance, taxes, grazing expenses = $58,700

We would have $58,700 to pay towards a $118,800 interest payment. We made almost half of the interest payment. You would have to double the stocking rate to 1400 stockers, two stockers per acre and assume you're going to get the same gain, (you're not) just to make the interest payment. There's no money left for a principle payment. I know I wouldn't be happy. A banker probably wouldn't be either.

Advantages of Leasing Vs Owning Land

Now let's look at leasing the 1100 acres. In my area land leases from $10-$20 per acre, depending on location and facilities. We will assume the highest rate of $20.00 per acre for this example:

1. 1100 acres x $20.00 = $22,000
2. We will assume the same stocking rate of 700 and yearly gain of 300 lbs.
3. We will assume the same grazing fee of $0.32.
4. We will assume same grazing costs.
5. Take $67,200 - $25,500 (lease and grazing costs) = $41,700 profit.
6. We made $59.57 profit per stocker by custom grazing.

Now we can make a cash deposit into the savings account instead of a payment to the bank. I like this scenario a lot better. I'm getting the reward from my labor, not the bank. By placing livestock ownership in front of land ownership you will be better positioned to pay off a farm if you decide to buy one. Young livestock eating tender vegetative grass is appreciating in value; this is where you want to concentrate on investing your equity.

Chapter 4

Tips For Finding Idle Land to Lease

To find idle land for lease, the first thing that is required is a local aerial map and county plat map from your local extension office. Look for undeveloped areas that have an adequate amount of open area. An existing good grass stand is a nice bonus. The bigger the area the more cost effective it will be to develop into a MiG system. You simply have more acres to spread your fence and water costs over. The income potential is also larger. Here is a list of items that I look for. Following each item is the reason why:

1. At least 60% open pasture must be available to develop into grazing.
Reason: Must have adequate grazing to cover costs and produce a profit each year.

2. Some source of water: pond, creek, well, or access to rural water line.
Reason: You have to have water. The more the better.

3. Electric pole, or existing meter a plus.

Reason: I do not like battery electric fence chargers. These electric chargers do not use much electricity. I've never paid over the minimum usage that is allowed each month.

4. Prefer gravel road frontage instead of paved road frontage.
Reason: There's a good chance that land will be cheaper to lease and if you have to move cattle across the road to the other side of the farm, you don't have as much traffic.

Start by drawing a five-mile circle around your home.

5. Farms bordering dead end gravel roads.
Reason: No traffic and private access.

6. Good fences.
Reason: This is a real work saver in getting the place animal tight, but these farms will usually cost you more to lease.

7. Abandoned neglected places.
Reason: These are very economical to lease, because there are usually no inputs in place.

8. A residence close to the area is a bonus.
Reason: Helps prevent vandalism to the property.

9. Retired or elderly landowners.
Reason: They don't want to sell their farm. They would like to see it taken care of.

10. Wealthy landowners.
Reason: They don't care about making any income from the property. They would just like their place to look pretty.

11. Land that is owned by people living out of state.
Reason: Absentee landowners like to know that somebody is looking after their place and taking good care of it.

12. Land that is close to my residence.
Reason: Less time and money spent on gas driving to it.

13. Problems or eyesores on the land.
Reason: These give me something that I can offer the landowner – I will fix these eyesores by correct management.

Example Of a Bad Land Lease Prospect

I had a tip from a neighbor that this new young landowner had bought a 630 acre farm about three miles from my house. It had been owned for 30 years by a doctor. The pastures had been severely neglected with hedge, cedar and thorn trees all over the pastures. This neighbor told me that he was looking for someone to lease the grass to.

Immediately I called him and asked if I could look the place over. I spent about four hours walking the perimeter fence, that is, what I could find of it. The only fence that would hold a cow was the fence along the blacktop. The other three sides were basically gone. The place was 60% woods, 10% open pasture, 30% grown up pasture.

I calculated that the farm had enough grazeable acres to support 30 cows, or 60 stockers if it was paddocked off. I called the landowner back and told him that there was not enough open land to support the number of livestock that I like to run. That was my polite way of saying no. Also, the massive brush clearing and fencing job that waited was not attractive to me. I don't mind cutting brush and building fence, as long as I'm enclosing some nice grazeable pasture.

That was three years ago, and the farm is still sitting vacant and getting grown up worse every year. These are the kinds of farms you want to avoid. They take tons of work to get them fenced. Then you don't have anything for all of your work when you do get them fenced.

Too Far From Home

I had a doctor call me one day about managing his farm for him. He found out from someone that I did MiG and he was convinced that I knew how to set up a MiG system. I must have impressed upon him that I knew what I was doing by talking to him over the phone, because he offered me his 160 acre grass farm to manage right over the phone. The only problem was his farm was about 45 miles one way from my residence.

At the time, I had just taken on a new farm and was in the middle of getting it set up for stockers. I also had seven other farms to look after besides working 40 hours a week in town. I had to turn him down, just because of the time it would take to drive there to run the farm. The doctor assured me it would be the cheapest lease I had, if I agreed to take it. I just plain hate driving long distances, especially when I have all the land that I need close to my residence.

For me this farm did not fit into my operation. It, and others like it, may fit into your grazing operation. It all depends on how far you decide to venture out from your home base. I like to be able to drive five minutes, jump out, move some calves and go on to the next farm. I have found that the more time spent moving calves means more money in your pocket also. They like to move, so get out there and move them.

Here is a list of items that I avoid. Following each item is the reason why:

1. Lots of houses close by.
Reason: Some urban people who move to the country don't like cattle in their backyard. They complain about the smell, noise, etc.

2. Lots of house dogs running loose in the area.
Reason: Several house dogs together will run and harass a group of cattle. I'm putting weight on calves, not trying to exercise it off.

3. Land with everything in place.
Reason: This lease will be expensive. Everything is already there. It doesn't leave you any bargaining power.

4. Land with minimal open pasture on it.
Reason: Will not be cost effective to graze.

5. Land that is farther than 20 miles from my residence.
Reason: Gas is expensive, so keep your farms as close to your residence as possible.

6. Unpleasant landowners.
Reason: I don't need them. There are too many nice ones out there. I get a huge reward from seeing a happy satisfied landowner. You can not put a money figure on that.

7. Leases less than 5 years.
Reason: If I decide to develop a farm, I want enough time to produce some nice profits.

8. No access to electricity.
Reason: I prefer plug in 110 volt fencers over battery type fencers. I don't have to worry whether my fence is hot or not.

Physically Finding Land

I like to start out by drawing a five-mile circle around my house. This means my house is the pivot point. I will drive five miles in any direction from my house to find and obtain leased land. This is very important when you are first starting out. Keep it close to your home. The reason for this is you don't need to spend a lot of time and gas driving to get to the land, which leaves you more time to work on developing it.

The land can have some brush and timber on it. I actually prefer some, because it makes it more wildlife friendly. Preferably 50-80% of the land needs to be open enough to grow some grass. I personally like to look for ground that has no fence on it, but is being hayed every year. These are gold mines, because usually there is an adequate stand of grass on the field and usually no sprouts because of being hayed yearly.

Lots of these particular fields are owned by absentee landowners who just give the hay to someone to keep their field mowed off and looking pretty. They don't have any other choice, because there is no fence or water to run livestock. I have talked to landowners who are actually exhausted every year, because they have to find somebody to mow their land off and give them the hay to do it. This is where our opportunity exists as graziers.

Drive around your immediate neighborhood with your

map looking at all the open undeveloped areas. You can get the landowner's name from the plat map. Get together the whole list of properties and owners' names. Decide on which property looks the most opportunistic for you. That is the one you will pursue first.

One thing you want to remember when looking at land candidates, the more attractive the property, which means good fences, blacktop road frontage, good water, corral, barn, etc, the more you are going to have to pay for the lease. You have less bargaining power because everything is already there.

As you get more experienced you may want to extend your range to include land that could be leased on your route to work, if you still have an off farm job. This way you could check stock either before or after work. This method opens up a lot more available area.

I am constantly seeing good pasture land sold off and developed into 20 acre house sites for people who want to live in the country. Well guess what? Those areas don't look like country anymore. You cannot blame the retired landowner if someone offers him $3000 per acre for land that he gave $100 for 50 years ago.

The Retired Landowner

There are landowners who are retired and have a nice nest egg saved who refuse to sell their land. They have lived their whole life on this piece of land, raised their family there, cleared the brush, run some livestock, maybe even did some grain cropping. They will not sell it for any price; it is part of their heritage. It would be like cutting their arm or foot off if they sold it.

If we young graziers could show these retired landowners what we can do with their farm, maybe we could slow down the mass movement of our rural landscape turning into housing developments. A lot of the elderly landowners want to see their land well cared for, for example, brush kept down that they spent a life time fighting. Approach elderly landowners who still have some livestock and offer to run their farm for them or as some kind of partnership.

This is a good way to get your feet wet without any risk. These guys are not stupid. Show them a working MiG farm that is set up; the diverse forages, fat steers grazing, clean water source, etc. I had one older gentleman farmer tell me when I started a new lease on the farm next to his, "Son I believe you are wasting your time trying to get clover growing on that old farm. I've tried and it cannot be done. You may get a little bit to come up, but the cattle and hot summer will just kill it."

Show the landowners some respect and listen to their concerns.

I just smiled and said, " Well I'm going to give it a shot anyway." I put in paddocks, applied lime, P&K and broadcast clover.

By the second year I had the most beautiful stand of red clover you ever saw on the whole farm. The old farm had never been fed any groceries and was continuously grazed to boot. The old gentleman farmer is completely in awe of how the place has improved. He is a believer in MiG now.

Show the retired landowner some respect and listen intently to his concerns. The majority of the land and cattle in the United States are owned by people over 60 years of age. A lot of these people would love to have someone they can trust to manage their land or operation. Go out and ask some of your local elderly cattle owners if you could custom graze some dry cows for them. This is a great way of getting your foot in the door. A dry cow has less nutrient requirements and is great to learn MiG with. Do your best with that set of cows and the next year maybe he will let you graze some calves. Later on, after the elder rancher trusts your management, ask about managing his cow herd or farm.

Grass farmers who already have a MiG system in place have a huge advantage over the novice just getting started. They can offer to show the landowner their own grazing system in operation, and explain to them that this is what their farm could look like in the future.

It is important to explain to the elder rancher the whole concept behind MiG: better nutrient recycling, catching more rainwater, keeping woody sprouts down, how the dung beetle comes right in behind the cattle and recycles the manure piles, how the earthworm population explodes adding humus to the soil profile with each cattle rotation, less manure runoff, diverse plant species, actually working in sync with nature, and being more wildlife friendly. You're actually building their soil up instead of robbing it by removing hay every year.

Attend as many Management-intensive Grazing courses and seminars as you can. Read the *Stockman Grass Farmer* and any other grass related topics you can. Never be satisfied that you have learned everything there is to grass farming. Always keep your mind open for new and better ways of doing things; this will increase your success.

Chapter 5
Show Them What You Got!

I brought a prospective landowner over to show him a working leased farm that had stockers running on it. It was around the end of May. Of course everything looked beautiful, with lots of legumes, ponds fenced off, good fences, a mob of fat steers grazing across the hill. Right across the road from my leased farm was an abandoned neglected farm. There were small cedars, thorn trees, sumac groves, and broomsedge everywhere.

I explained to the landowner that my leased farm had looked just like that two years earlier. I had bushhogged the cedars and thorn trees, and the cattle did the rest. The owner could hardly believe that the land looked so much better in only two years. I asked him if I could move the stockers real quick. He stated that he had recently gone through a knee surgery and was not up to walking down the hill to herd them to the next paddock.

About that time I let out a war whoop. 300 yards away, 100 steers came busting down the hill, bucking, twisting in the air, bellowing and charging right up to the gate. I stepped over the fence and opened the gate and stepped back. The whole herd charged through in 30 seconds and were just flat grazing up a storm on the new grass and legume paddock. All you could hear was grass being snipped off by the steer tongues. I couldn't have rehearsed a more awesome display of MiG in action!

As I closed the gap the older gentleman's mouth was wide open in astonishment. He remarked, "I'm 78 years old and never in my life have I seen anything like that. How did you get them to do that? Those calves must have been somebody's pets?" I assured him that they were not pets, but were trained by the promise of a new salad bar every time they heard me holler for them. The gentleman gave me the lease on his land.

I was interested in leasing some land that was owned by an out-of-state landowner. I was able to track down her name and called information to get her phone number. I introduced myself over the phone, and told her where I lived in relation to her farm. I asked her if she would be interested in leasing her farm for grazing? She was very interested and was kind of frustrated with trying to find somebody to bale the hay off of it every year.

I explained that if I leased it, it would be rotationally grazed between four paddocks that I would install. I also sent her one of my narrated videos of a paddock move. Included in the video were some water systems that I had developed along with the lush pastures of grasses and legumes. I asked her what kind of income she was expecting from her property? She replied that she was presently getting $370.00 a year for the hay crop.

She stated that the hay bales were getting fewer in number every year, along with it the income. I got a five year lease on the 40-acre solid-grass farm for $400 per year. This same lady remarked after the third year of the lease that while she had owned the property for 15 years that the past three were the most rewarding. She went on to say, "Greg, you are doing such a good job of improving the looks of my farm, that I feel actually connected to it. I actually feel like it is a productive working farm, instead of a dull piece of ground." I got all kinds of satisfaction and reward from that comment.

Keep Your Farms Neat

It is important that if you have a working MiG system, you make sure it is neat before you bring over a prospective landowner. Don't leave any old implements lying around or any trash, just have pretty pastures of grass and livestock to show. I'm not saying that all the fences and buildings have to be painted, but your place should not have a cluttered look. The prospective landowner is going to give your place the eyeball. If it is trashy looking, then you can forget about leasing his land. I don't blame him. I wouldn't want my farm looking trashy either. You want to impress upon the landowner that you care about his land and his land may well look like this someday. I'm convinced there is no stronger selling point than a pretty MiG farm. Your neat and clean farm tells the landowner that you care about yourself also and have pride in what you do.

Showing Your Farm and Walking Theirs

When you bring the landowner over, show them the legumes, the diverse plant life you have in your paddocks, the mulch accumulation you have below the growing canopy to catch all the rainwater so it doesn't just run off your property. Point out how the manure piles with proper MiG rotation melt into the ground after a good rain.

Show them an ungrazed paddock versus a grazed paddock. These are the kinds of things that will impress upon the landowner your intent for their land. Explain to them how environmentally stimulating MiG is to the land. You need to explain to them that their land is one big solar collector and that it is your goal to have as many green growing leaves (solar collectors) as possible, exposed to collect this energy.

I showed my grazing system to a landowner who did not know the difference between grass and a legume. I took it slow and showed him my diverse pastures of legumes and cool-season grasses. I took out my pocket knife and cut out a wedge of sod and showed him how moist it was even though it hadn't rained for two weeks. But guess what? After explaining the total system to him he could hardly wait for me to get started on his land.

I like to show a prospective landowner one of my working MiG farms that has about the same topography as their farm. This makes it easy to show them what parts of their farm will look like in several years.

When you walk their land with them, point out the lack of plant diversity and how you can correct it with proper grazing management. Emphasize to them the importance of giving

Showing a Landowner Your Farm

** Farm should be neat, no trash, just healthy green pastures with cattle grazing.*

** Do a paddock change.*

** Point out a grazed paddock versus an ungrazed paddock.*

** Show them a fenced pond with cattle drinking behind it, not in the pond.*

** Explain that with proper grazing management their farm could look like a MiG farm.*

grass a rest period, which makes it grow back faster. Emphasize to them that a grass plant needs to be grazed. That is what keeps a grass pasture healthy and the weeds out, if the grazing is managed correctly.

While you are walking the landowner's farm with them, do not ever climb over a fence. Either slide under the fence or find a gate. Nothing makes a seasoned landowner madder than to hear their wires being stretched. I will pick up loose cans or bottles while we are walking the property. It impresses upon the

Walking the Prospective Landowners' Land

* *Point out the strong points of the idle land.*
* *Point out the problem spots on idle land.*
* *Explain how you can improve the problem areas with proper grazing management.*
* *Ask what their long-term plans are for the land.*
* *Ask what the maximum length of time for which they would be interested in leasing the land.*
* *Offer to write a proposed lease contract for them to look over.*

landowner that you are not a litterbug.

I had one landowner who was very concerned with having all of his land covered with big old cake-size cow pies that he would have to constantly be on the look out for while walking his property. I explained to him that with high stocking density how the cattle mob mowed the pastures, which kept the forage tender. The calves' stomachs digested the grass very quickly and the manure piles looked more like small thin saucers.

I explained to him that by the time the cattle were back to that paddock you could hardly tell where they had been. Between the dung beetles and the rain, they just melted away and disappeared leaving behind the wonderful soil building nutrients.

Explain to the landowner that 90% of their land is being rested at all times. This allows wildlife to take advantage of all the succulent forages. It actually stimulates wildlife production because more beneficial plants are being grown.

Ask the landowner what their long-term plans are for the property and what their biggest concerns are with the property. Walk their land with them pointing out how you would run your paddocks. Have an aerial map of their land with all proposed paddock divisions drawn on it. Show them how the livestock will be moved from one paddock to the next.

After you have explained the total concept behind MiG, ask them what would be the longest period they would be interested in leasing the land to you. The absolute minimum is five years, ideally ten years is wonderful. You need to explain to them that it may take three to four years to turn their land around.

Chapter 6

Calculating the Land Lease Contract

Before you negotiate a price, estimate how many cows, stockers, etc, you will be able to run on the land once you get it fenced for MiG.

I prefer stockers. They are easier on the ground. If you get caught in a rainy period they will not pug up the pastures as severely as 1200-pound cows. If the grass is decent, there is more money to be made by putting weight on a light calf than grazing cows.

Here is an example of how I calculate potential gross income and the costs of setting up a grazing system on a prospective farm that has 80 acres of fair grass, no legumes:

Notice I said grass, not brush or timber.

The 80 acre farm measures 2640 feet long by 1320 feet wide. The farm has had no fertility or lime added to it. The landowner is not interested in investing any more money into the property. What you see is what you get! The perimeter fence is okay, but there are no interior fences on the farm. It has one pond on each end of the property.

I will run one hot wire around each pond and run 1" polyethylene hose over the dam with a siphon. I will have a 400-gallon stock tank with a float valve behind each pond dam. I know that when the weather gets hot, the grass growth will slow down. I will need to plan a rotation system that gives the grass 30 - 40 days of rest. Each pond will service four paddocks. This gives us eight, ten-acre paddocks. Each paddock will be a single electric hi-tensile wire division. I will be grazing stockers in this example.

Calculating Gross Income Potential

1. We will graze a group in the spring from April 1st until August 30th, which gives us 150 days of grazing for the spring herd.

2. The custom grazing fee that we are charging the cattle owner is $0.35 per pound gained. The stocker owner supplies all inputs: mineral, salt, medicine.

3. We will assume an average daily gain of 1.3 pounds per stocker.

4. We'll assume a stocking rate of one stocker per acre.

5. For the spring herd take 80 stockers x 150 days x 1.3 pounds gain per day = 15,600 pounds of custom gain.

6. Take 15,600 lbs x $0.35 custom grazing charge = $5460 gross income for the spring herd.

7. Let the pasture rest 45 days and graze another group from October 15th until December 30th. That gives us 75 days of grazing for the fall herd.

8. For the fall group take 80 stockers x 75 days x 1.3 pounds a day gain = 7800 pounds of custom gain.

9. Take 7800 pounds x $0.35 per pound of custom grazing charge = $2730 gross income for the fall herd.

10. The total gross income for the spring and fall grazing herds is $8190.

11. The total pounds of gain from both herds is 23,400 pounds.

Costs of Implementing a Grazing System

1. Two used stock tanks, $50.00 each = $100.00
2. Two stock tank floats, $9.00 each = $18.00
3. Two 100' rolls of 1" polyethylene pipe, $16.00 each = $32.00
4. High voltage fencer = $130.00
5. Two 4000' rolls of hi-tensile wire, $68.00 each (see note below) = $136.00
6. 133 - 1/2" fiberglass lineposts, $1.80 each = $239.00
7. Seven boxes of wire crimps = $10.00
8. 12 ratchet tensioners, $1.00 each = $12.00
9. One roll of white poly tape for gates = $22.00 roll
10. Electric meter charge, $8.00 per month x 12 = $96.00
11. 16 hours labor installing system, $20.00 per hour = $320.00

Total = $1115.00

Tip: Ask your local electrician to save his insulated wire scraps for you. They make great electric fence handles, and are free. If you run over the insulated wire handle you will not have the worry of breaking it. I hate to think of how many plastic gate handles I broke driving over them.

Now we will deduct the cost of implementing the grazing system from the potential gross income that the 80 acres will generate:

* Potential gross income = $8190.00 - $1115.00 (fence and water costs) = $7075.00 (Gross income, lease amount not subtracted yet)

What Is the Pasture Worth?

Naturally you want to lease the farm as economically as possible. You have calculated a rough estimate of the potential profit that the farm may generate and the costs of implementing a grazing system on the farm. Now how much are we going to pay for the lease?

First of all, you never make an offer. It's his land. By making an offer, you may offend him if it's too low. You ask the landowner how much he would lease the farm to you for?

Every lease is different; every landowner is different. Some landowners don't care about making any income, but they are extremely concerned with keeping their property looking good. Another way of calculating how much you can afford to pay is to decide what your labor is worth.

Let's assume we put a value on our labor of $20.00 per hour. Let's assume we spend one hour per grazing day moving and checking the calves. 1 hour x 225 days = 225 hours labor. Take 225 hours x $20.00 per hour = $4500.00. We have a rough calculation of $7075 gross left after implementation costs. Take $7075 - $4500 (labor for moving and checking stock) = $2575 available to offer for lease. Take $2575 divided by 80 acres = $32 per acre is the maximum you can pay for the lease and still make $20.00 per hour for your labor.

By having to install the grazing system, I would never pay over $15.00 per acre for the first year. Assuming we leased it for $15.00 per acre, that would be $1200 per year for the 80 acres. You're earning $26 per hour for your labor. This gives you some breathing room for your establishment year and some money to put toward some lime, P&K. Adding back some improvements, there is a good chance you can increase your stocking rate from 1 to 1.5 stockers per acre on the second year of the lease.

Now you're looking at stocking 120 calves in the second year of the lease.

It still absolutely amazes me how an old worn out farm can explode with good forage, with just some simple nutrients added along with proper grazing management.

For the first year of grazing, always calculate on the conservative side. Your forages and daily gains should improve each year with proper management. Remember, the landowner is going to get his biggest financial gain (increase in land value) from what you are doing to his land, if you manage it correctly. To me there is nothing more satisfying than taking a piece of marginal land and turning it into a grass-grazing haven.

Explain to the landowner that you will be investing lots of labor and resources to develop his land into a grazing enterprise.

Note: Wire was roughly calculated by taking the measurements of the 80 acres. On a piece of paper draw a rectangle. Then draw a line down the center splitting the rectangle lengthwise. This line would measure 2640’, which is the length of the 80 acres. Then draw in your eight paddocks. It is easy to measure how much wire is required from the sketch on the next page. It comes out to 6600' for eight single wire paddock divisions. Add an additional 1400' to go around the two ponds. 8000' total wire for the entire 80 acres. For the line post calculation take 8000' divided by 60' = 133 line posts. Remember the posts and wire are yours. They are your biggest material cost of leasing, but you can use them again on the next farm.

Farm Sketch Explanation

F = Fence
G = Gate
L = Lane
P = Pond
T = Tank

Farm is 2640’ by 1320’
The eight paddocks each contain 10 acres. Total 80 acres.
Interior fences will use single wire.
Gates use single strand 1/2” white polytape.
Assuming the perimeter fence is functional, calculations for the wire and posts for interior paddock divisions would require 8000’ total hi-tensile wire
133 - four foot posts (60’ spacing)

(See diagram next page.)

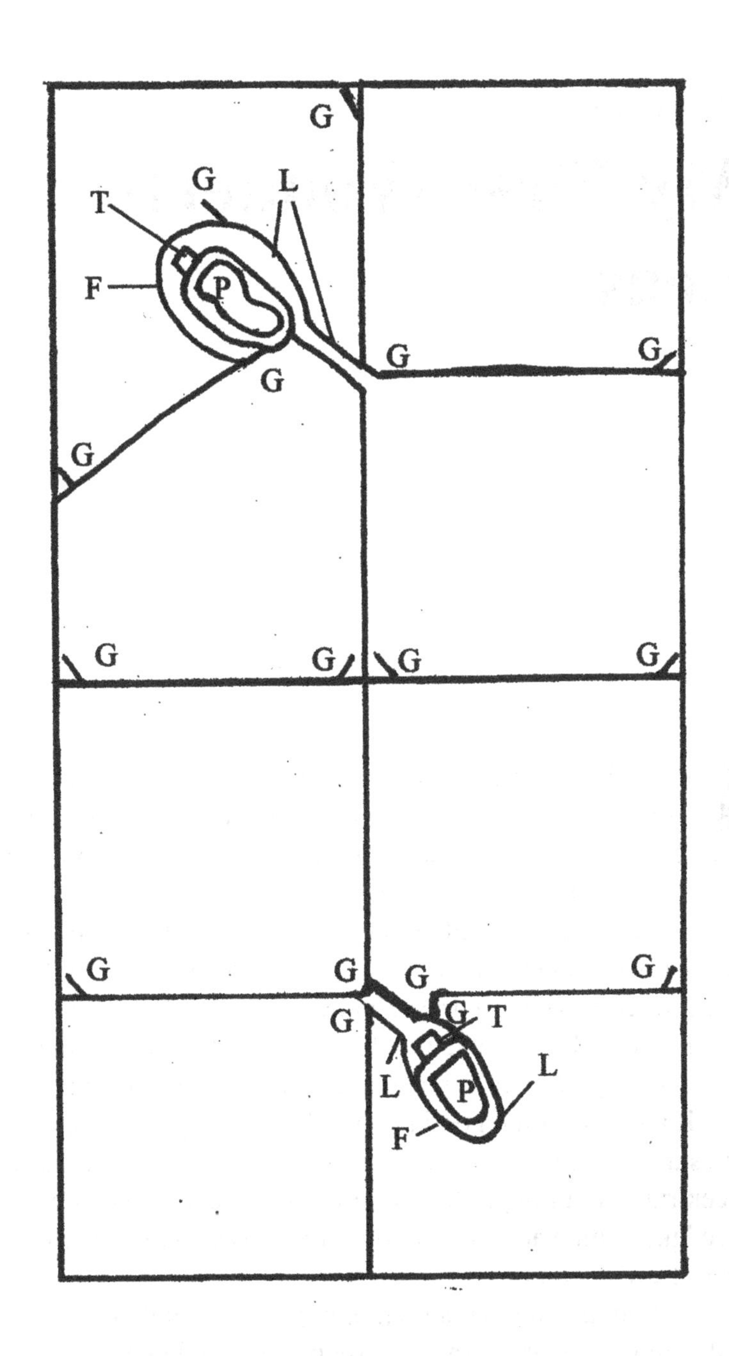

80 Acre Farm Sketch, 8 Ten-Acre Paddocks

Chapter 7
My First Wonderful Lease

I had driven by this vacant pasture land about two miles from my farm for many years. Never once did I look at it as a candidate to lease. Leasing was never an option in my mind. My mindset was you had to own it.

The farm was 150 acres; 60 to 70% of it was open land, the rest was timber and brush. The farm had a lot of problems. It was overgrown with brush, had very steep ridges, limited water, and had been hayed to death. The only fence on the farm was one strip of road frontage and a section between a neighbor. It also had some good potential: a pretty good base of grass under the small brush, no big brush in the pastures, a small creek running through the property, one decent pond, a one-acre lake, a nine acre bottom of gama grass, and it was two miles from my house.

I got the landowner's number from a neighbor and called to inquire about their property. The landowner was completely caught off guard by my call. I asked him if he would be interested in leasing his farm to me for setting up a grazing system. He told me that they had been looking around since they bought the property to see if they could find somebody to take care of it. They had shown the farm to two other local cattlemen and were turned down by both. The reason they were turned down was too much fence was needed.

They had bought the farm to hunt, fish and build a summer home on when they retired. They were getting very frustrated trying to manage the farm while living in another state. They would spend several weeks every year fighting the brush, and the next time they came up to the farm it was always grown right back up with brush. They were thinking of giving up the brush battle and selling the farm. Both the husband and wife had been raised on a farm, so they could understand the concept of MiG when I explained it to them.

The landowner stated right from the start that if they agreed to any form of grazing that they wanted all livestock kept out of their pond and lake. I assured them that the pond and lake would both be fenced off, excluding the livestock. I asked if they would mind if I walked the farm, then I would call them back. They agreed.

They were very interested in doing something with it, because they couldn't since they lived in another state. The owner explained that the back 40 had no fence period, just a survey stake.

As I walked the property, I noticed that the steeper areas of the farm had the best grass. These were the areas that were too steep to be hayed, therefore all the fertility had not been robbed. These same steep areas also had massive thickets of blackberry bushes, sumac groves, thorn trees and dewberry vines. Not exactly prime stocker calf forage.

Even with all the brush there was a decent stand of fescue nestled down in the dead thatch, but no presence of legumes due to the ground being covered by thatch. I remember the first time I fought my way down the hill, through the brush choked valley and crawled to the top of the big ridge that overlooks the whole farm. As I stood there getting my breath, I could visualize where the paddock divisions would go. I mentally removed the brush thickets and placed a large herd of stockers grazing in the valley below me.

By the time I reached the back 40, forage was getting better and better and my optimism was at it's peak. Then I realized there was no water to feed the back 40, even though it had a one-acre lake on it. The landowner had made it very clear that he did not want any cattle activity even close to the lake. When the lake was built, it had a 1.5" water pipe with a shut off valve in the bottom of the dam. The only way you could get to the back of his dam was to come off his neighbor's 70 acres or go right down the back of the dam itself. That was a sick feeling; all that grass and water, and absolutely no way to use the water.

Keep working toward your goal and things will get easier as time goes by.

The farm also had two nine-acre bottom fields of gama grass, a four-year-old strong stand. One gama bottom was in CRP, the other was available to graze. The east 30 acres bordered a blacktop and gravel road. There was a good six-strand barb wire fence between his neighbor and the landowner's land. The other three sides had no fence. There was no water on the east 30 acres either. The rural water line ran full length along the blacktop frontage, but there was no meter. The middle of the farm had one old farm pond and a small creek that ran down the middle. I initially dismissed the creek as a reliable water source, but later found out that it was spring fed.

I got an aerial map of the farm from the local NRCS. I drew out the property lines of the farm and shaded in the grassy open areas with a black marker. I asked an NRCS technician to measure the open areas. This gave me an accurate measure of available pasture. I came up with 90 grazeable acres on the farm. From the aerial map I was also able to calculate how many feet of perimeter fence was needed.

Calculating Costs of Implementing Grazing System

I used the aerial map to calculate total linear feet of the perimeter and paddock division fences. The farm consists of three full quarter sections and one half of another. A 1/16 section measures 1320' x 1320' (40 acres). The farm is shaped like an L, with the back forty making up the "foot " of the L. Here is how I calculated the total fence:

* Blacktop frontage perimeter 1320' x 5 wires = 6600'
* Gravel road frontage perimeter 1600' x 3 wires = 4800'
* Perimeter fence not touching road = 3960' x 2 wires = 7920'
* 10 paddock divisions 10,000' x 1 wire = 10,000'
* Total linear feet of all fence = 16,880'
* Total wire needed for all fence = 29,320'

Estimated Cost of Total Materials

1. Eight 4000' rolls of hi-tensile wire, $68.00 each	= $544
2. 281- 1/2" lineposts, 60 foot centers, $1.80 each	= $506
3. High voltage fencer	= $130
4. 3 boxes wire crimps, $10.00 box	= $30
5. 50 ratchet tensioners, $1.00 each	= $50
6. One 660' roll poly tape for gaps, $22.00 roll	= $22
7. Electric meter charge, $8 x 12 months	= $96
Total cost of grazing system	= $1378

Note: I already had corner posts, a stock tank and garden hose for the pond siphon. I did not calculate any labor for putting in the system. No corral was figured in, I planned to rent a portable.

Estimating Gross Income Potential

I calculated the profit potential of the 90 acres for the first three years. The existing forage base did not have the potential to run stockers on it, so I calculated custom grazing cows as a clean up crew the first year. Beginning with the second year, I planned to frost seed red clover, and graze stockers. Here are the calculations I came up with:

Year 1: March 1 to September 30 = 7 months grazing
1) 30 cows and bull x $10 per month = $310 per month
2) $310 x 7 months = $2170 gross income

Year 1 gross income = $2170

Year 2: Calculated custom grazing 85 steers in spring turnout
1) Graze April 1 to August 31 = 150 grazing days
2) 150 grazing days x 85 head x 1.25 lbs gain/day = 15,937 lbs of custom gain
3) 15,937 lbs of gain x 32 cents = $5100 gross income

Rest pastures 45 days and bring in 85 fall steers:
1) Graze from October 15 to December 15 = 60 grazing days
2) 60 grazing days x 85 head x 1.25 lbs gain /day = 6375 lbs of custom gain
3) 6775 lbs of gain x 32 cents = $2040 gross income

Year 2 gross income = $7140

Year 3: Calculated custom grazing 1.5 steers/acre = 135 steers
1) Graze April 1 to August 31 = 150 grazing days
2) 150 grazing days x 135 head x 1.35 lbs gain/day = 27,337 lbs of custom gain
3) 27,337 lbs gain x 32 cents = $8748 gross income

Rest pastures 45 days and bring in 135 fall steers:
1) October 15 to December 15 = 60 days grazing
2) 60 grazing days x 135 head x 1.35 lbs gain/day = 10,395 lbs of custom gain
3) 10,395 lbs gain x 32 cents = $3499 gross income

Year 3 gross income = $12,247

I figured the lease years four through ten would be at least equal or better than year three because the forage base would be stronger as time went along if I managed the land correctly.

I took the seven remaining years and multiplied them by $12,000 = $84,000 gross custom grazing income. All ten years totaled $105,500 gross income!

I was getting pretty excited by now about the potential this farm had. I now had the cost estimate of implementing the grazing system ($1378) and the potential gross payback for the ten-year lease was $105,500.

At that point I did not know what my net income would be, because I did not know what the lease was going to cost. I knew that there was a ton of work ahead of me if I decided to tackle the fenceless farm.

I called the landowners back and told them I would be interested in leasing it, only if they would give me a ten year lease.

They did not object to a ten-year lease proposal. They figured they were both at least ten years away from retirement anyway.

I also voiced my concern over not having any water on the back 40 and no water on the blacktop pasture. The landowner seemed interested in fixing the water deficiency problem if the first year lease went well. I figured I could have a small pond built on both sites if it came to that.

I explained to them that I was going to have a lot of hours wrapped up in labor just getting the farm perimeter tight so that it could be grazed. I offered to write up a contract for him and his wife to read over.

Listed below and on the following pages is the information in the contract proposal that I sent them. Each paragraph is numbered and followed by an explanation of why each sentence was put in the lease proposal.

150 Acre Land Lease Proposal

(Paragraph 1) 1. I, Greg Judy, agree to lease from the owners, John and Sally Doe, their 150-acre farm located in Boone County, section 10, township 8, for a period of ten years. 2. The lease will not start until the day the cattle are placed on the farm.

(Paragraph 1 breakdown): 1.You want to have your name first, then the owner's name, a legal description of their farm, the length period of lease. 2. When the lease starts.

(Paragraph 2) 1. Leasee is to provide all fencing materials and labor to install hi-tensile electric fence around the farm that is void of fence. 2. At the end of the lease, all posts and wire belong to leasee. 3. Landowners will give leasee 90 days to remove posts and wire from property.

(Paragraph 2 breakdown): 1.You are verifying all labor and fencing materials will be supplied by you. 2. You also have it in writing that the posts and wire are yours at the end of the lease. 3. This gives you time to remove the fence.

(Paragraph 3) 1. Leasee will initially start erecting 3 wire hi-tensile fence along gravel road, then proceed to fence off gama grass field separately. 2. Leasee will install paddock divisions that hook to pond tank. Pond will be fenced off, with siphon hose feeding water tank. 3. The CRP field will be fenced off to prevent grazing, along with the buffer strip along the creek. A buffer strip around the lake will be fenced off to prevent cattle from grazing this area.

(Paragraph 3 breakdown) 1. I am explaining the sequence that their farm will be developed. 2. I am explaining how the paddock divisions will get water. 3. This also covers the areas they had concern with – the pond, lake, CRP, buffer strip.

(Paragraph 4) 1. Cattle owner will have a liability policy in effect before the cattle are placed on the farm. 2. Cattle will be rotated through the paddocks allowing the grass a rest period to grow back. 3. Leasee agrees to broadcast 3 lbs red clover per acre each February to help improve the forage base. 4. Cattle will be placed in an area during deer season where they will not interfere with hunting. 5. Leasee agrees to control trespassing on the property.

(Paragraph 4 breakdown) 1. I am promising a liability policy to protect him from any accidents that may occur if a calf gets out. 2. This assures him also that his land will not be overgrazed. 3. I will put down red clover each spring. It will help his wildlife and my stocker gains. 4. He was concerned about the possible conflict of deer hunting season with the cattle. 5. The landowner was wondering if I would post the property and control trespassing.

(Paragraph 5) 1. Leasee has 100% control over all forage and grazing management decisions. 2. Leasee agrees to assist his labor in any water or pasture improvements that the landowner undertakes. 3. In the case of a drought, to where there is not any forage to graze, no cattle will be grazed that year. The drought year will not count as a lease year, nor will any lease payment be due. 4. Lease payment will be due twice yearly, November 1st and May 1st.

(Paragraph 5 breakdown) 1. This sentence states that I have 100 % control, which gives me maximum flexibility. 2. I had stated earlier that if he ever wanted to tackle any improvements that I would supply the labor if he supplied the finances. 3. I was really concerned with facing a drought, not having any forage to graze, yet still faced with coming up with a lease payment! 4. Gives the actual dates that the lease payments are due. I timed the land lease payments to come due when there were no other big bills hitting me.

(Paragraph 6) 1. Proposed lease payment on 150 acres:

Year 1 = $300
Year 2 = $300
Year 3 = $350
Year 4 = $350
Year 5 = $400
Year 6 = $400
Year 7 = $450
Year 8 = $450
Year 9 = $500
Year 10 = $500

2. Leasee Signature______________ Date____________
Landowner Signature __________ Date_____________

3. Note: John and Sally Doe list any changes or concerns they may have with the above contract. This is a rough draft, but I think it covers most of your major concerns.

4. Yours truly,

Greg Judy

Date____________

(Paragraph 6 breakdown) 1. This is the complete time schedule of proposed lease payments for the ten years. I knew I had to cover my unknown labor costs and they were going to be high. Their present income off the hay crop was between $150 - $200 a year. The landowners also mentioned that they did not buy the property to make an income off of it. They were extremely concerned with having it taken well care of and improved upon, from what its present condition was. 2. Both of our signatures and date. 3. This encourages them to bring up any concerns that I left off. 4. My signature and the date.

They signed the contract in January and I was ready to go to work immediately. All winter long I had been cutting, drilling, and painting scrap fiberglass rods that I bought as factory rejects. The first thing I tackled was the perimeter fence around the land that had water. I strung three wires along the gravel road and two wires on the rest of the perimeter fence that did not touch the road. I fenced off the pond and lake. Realizing that the little creek was spring fed, it never went dry the whole summer, even during the drought.

I went into the local sale barn and posted an ad looking for 30 cows to custom graze starting March 1st. I asked the lady behind the sales counter if she minded that I posted a grazing ad? She said it was okay, then followed me over to the board to read it. She asked me if she and her husband could come out and look at the farm, since she was very interested in grazing some cows. Their cows were registered Charolais that would have 4- to 8-week-old calves by their sides.

I did not know at the time that she was the wife of a family member who owned the sale barn. I knew the family was honest, hard working and well respected in the community. Before I could get out of the sale barn, a veterinarian told me if I came up with any more pasture to graze to give him a call. The cattle owners came out and I showed them the farm. The farm looked pretty rough, but had a pretty good stockpile of grass in the open areas. I only had it partially fenced, but explained to them where the paddock divisions would go.

I explained how the cows would be moved from paddock to paddock, giving the grass a chance to rest. I told them about the gama grass, how it really kicks in growing when the other cool-season grasses slow down. I promised to do my very best to keep their cows in grass and watch for any health problems. I told them that I would like to have a written contract as this would prevent any confusion. They were strongly in agreement with me on having a written contract. I agreed to write a custom grazing contract for them to proof read.

Listed below is the grazing contract that I wrote:

John And Sally Doe Grazing Contract

John and Sally Doe agree to pay Greg Judy $10.00/month per cow-calf pair. They will stock 30 cow-calf pairs and one bull. Pasture rent is due at the end of each month. Greg Judy is responsible for moving cattle as needed, depending on grass growing conditions.

Cattle will be checked every two days. If any health problems are noticed, Greg Judy will notify cattle owners. John and Sally Doe are to provide the salt and mineral. Greg Judy will be responsible for keeping mineral mix with cattle on each paddock change.

John and Sally Doe take full responsibility for any death, injury, or loss of cattle. John and Sally Doe agree to have a livestock liability policy in effect before cattle are placed on the farm.

Cattle will be grazed from March 1st to September 1st. If a drought hits, Greg Judy will give cattle owners 30 days' notice to remove the cattle.

Grazier ____________________ Date______
Cattle Owner________________ Date______

The cattle owners signed the contract in February. Now I had until March 1st to finish up the fencing of the farm. I used my coon hunting light to build fence at night after I got off work in town. It was always dark when I got off work in the winter, so I just strapped on my coon light and went to work. It works great. Your hands are free and you can see just like it is daytime. The light is mounted to a hardhat on your head. It is re-chargeable up to 500 times. I found it to be a great stress reliever from working in town all day.

Usually I worked two to four hours each night cutting brush, driving posts, stringing wire. It was great. It sure beat sitting inside all winter watching worthless TV every night. Nobody bothers you, you get a lot of work done. Usually there is no wind and the weather is very mild as long as you're working. I got some pretty weird looks from people going by. My girlfriend (now my wife) helped me every weekend building fence. Talk about testing a relationship!

The Rain Cometh!

30 cow-calf pairs were delivered the first week of March. The cows had been on hay all winter and were ready to do some grazing. I had about 50 of the 90 acres fenced. Starting in mid April, it rained 2" to 3" every day for three weeks straight. I was ready to start building a boat, because it looked like it would never stop raining. All the fields went to mush, so I kept the cows up on the highest ground, but ran out of grass pretty shortly.

When it seemed like things couldn't get any worse, they did. I came by to check the cows after work one night. It was raining like it had every day that month. The cattle owner was stomping up the hill toward me in a downpour. He was madder than a hornet, because the neighbor's mongrel bull had decided to come across the fence to visit the new ladies. He explained that when he pulled up to the gate, there was Rufus mounted on his favorite cow.

Rufus had a head on him like a buffalo and a body like a sickly calf. You couldn't draw a picture of a poorer specimen of cattle genetics. I quickly assured the cattle owner that Rufus would be off the farm by dark, even if I had to drag him off stone dead. I worked three hours getting him out of brush thickets, but finally got him back to his owner. We put him in his barn for the last time. Rufus was sold the next morning. The cattle owner was able to give the cow a shot of Lutalyse to prevent a baby Rufus the next spring.

I let the cows graze the rest of the paddocks, but moved them every couple days. This helped the pugged areas recover some between rains. They couldn't help but pug wherever they walked. The ground was like a big sponge. I didn't realize how soft an idle, heavy thatch, thick mulch forage stand could be. The mulch soaked up and held all of the rain, which made drying out almost impossible. The farm had not been grazed for 15 years, which added to the soft ground.

I was dreading the idea of having to explain the terrible pugging to the landowners. One bright spot though, was that the cows were eating everything that wasn't a cedar, or didn't have a thorn on it. I called the landowners up, told them about the pugging, also mentioned the brush removal the cows were doing. I burned off the gama grass field and put down 80 lbs of N per acre. The gama field was grazed one week each month until September.

The landowners came up in May and were amazed we had put up so much fence. They were also surprised at the absence of brush thickets, complements of the cows. They were very pleased with what the farm looked like, except of course for the pugging. The absence of the brush and the pretty fences helped smooth the pugging issue. I told them that a light disking with a harrow in tow, would smooth out most of the holes.

The landowners offered to pay for having a meter set and put in a hydrant on the 30-acre blacktop frontage pasture.

They also wanted to do a cost/share pasture improvement program. The cost/share pays 75% of all lime, fertilizer, no-till drill rental and seed. I did all the paper work for them, and was approved for a fall inter-seeding on the entire farm. Next, I got the meter with hydrant attached and set on the blacktop frontage pasture. This gave me 30 more acres for the cows to clean up. I also fenced the back 40 and made the cows walk back to the creek that ran through the center of the farm. It was not the best, but it worked and the cows did not seem to mind the walk.

The Drought

Starting in June, we did not have any measurable rain the rest of the year. The gama grass was kicking in though. It liked the hot weather, and actually grew faster. There were cracks in the ground between the gama grass that you could drop a golf ball in, but it kept on growing anyway. Some of the cows were first calf heifers, and the gama grass really put the weight back on them, despite the drought and suckling a calf. It's a wonderful feeling to be able to turn a group of cows into a thick lush green gama field when everything else is brown. No wonder they call it the "Ice Cream Grass Of Grasses."

The landowner offered to build two ponds – one right in the center of the back 40, the other one on the front of the farm. It would feed three hill paddocks and the gama grass bottom. I did all the plumbing work, seeded, fertilized and strawed down the entire excavated area of both ponds to prevent them from eroding during rains. The cattle owners came and got their cows the first of September. They were very pleased with the condition of them. Despite the drought, I had managed to put some flesh on them by rotating them every couple of days.

Electric Fence Corral

I did not have the money to invest in a standard corral system. So I set four posts and strung five hi-tensile wires with a ratchet tensioner on each wire. The corral is 80' x 80', which took 1600' of hi-tensile wire to build. I tied white ribbons to each wire so that the cows could see them. The cattle had been trained that white ribbons meant "Pain." All my paddock division gates are white electric poly tape. The cows had a good sniff of these throughout the summer.

The corral loading alleyway was made from free steel factory siding that locked together, is very strong and cattle can not see through the walls of the alley. The steel siding was fastened to cedar posts cut from off of the farm.

The total purchased materials for the corral was 1600' of

hi-tensile wire and five ratchets; everything else was free. When the cattle owner pulled up the day of the loading, he had his doubts that the corral would hold anything. To be perfectly honest I also had some doubts. I had a one-acre catch pen hooked onto the entrance to the corral. It was made up of two hi-tensile wires. The haulers, cattle owner and I walked the cow-calf pairs into the hot wire corral with a white ribbon held between us. We sorted the calves out for one load and the cows were loaded separately.

The key to obtaining great leases is fully explaining MiG to the landowner.

Even with all 30 cow-calf pairs and a bull pushing on each other, they did not mess with my hot wire corral. When they did touch it, all you heard was what sounded like a .22 rifle going off, "ker-pow!" They had a high incentive to get off of it. All the rest of the farm was shut off from the electric fencer, except the corral. The wire on the corral was reading 7200 volts, so I'm glad that I didn't get bumped into it.

That first summer of grazing the leased farm taught me a lot. With heavy spring rains came pugging, then the neighbor's rogue bull, managing the grass through the summer drought, having limited watering areas; but through it all, I kept right on doing the best that I could with my limited resources. Things worked out. The cattle owners were happy with the condition of the cows. The landowners were happy with the improvements that I had made on their farm. Tough times make you appreciate the good times. The lesson I learned is to keep working toward your goal and things will get easier as time goes along.

Lime and fertilizer had been spread on the entire farm in July. Most of it was still right where it had landed. The night dews kind of melted some of it into the ground surface, but the dry cracks in the ground were full of it. I drilled red clover on the entire farm in September, and hoped that our usual fall rains would come. They didn't.

What About the Land Across the Fence?

About that time I noticed that nothing was being done with the 70-acre tract of land across the fence from the 150 that I had leased. It did not have any useable water on the property. But I already had a plan to fix that. I did not know who owned it, but after checking I found out the owner lived out of state. I got his phone number and called him about leasing his vacant land. I already had part of his 70-acre tract fenced, due to fencing the 150-acre farm that I had already leased.

I met with the retired landowner of the 70 acres. The landowner also explained right from the start that he was not expecting to make a lot of income from the lease. He explained that he wanted the land to be taken care of and look nice. He wanted the land to be kept clear of brush and have the mowed-grass look. He also had a new lake, which he was concerned that the cattle might destroy. I put his fears to ease by explaining that the lake would be fenced off to keep the cattle out. I showed him my other farm next to him and explained MiG. I told him how clovers have the unique ability to fixate nitrogen from the air and store it in their roots, then the grass roots come along and steal it. He had never heard of this and was intrigued.

I showed him how a mob of cattle had healed erosion ditches by sloughing off the edges and adding manure for fertility. Then I explained the importance of resting the grass and leaving a solar collector intact when I moved to the next paddock. All he had to do was look over the fence and see what was happening on my existing leased farm and compare it to his farm. I negotiated a very economical six-year lease from the landowner. The first year was free, the remaining 5 years I would pay $200 a year for the 70 acres, which came out to $2.86 per acre.

Let me make something very clear though. These kinds of leases are possible only after you have taken the time to completely explain the whole MiG concept. Once you get people excited about what you're doing, they want to be part of it. All people hear about on the news is how bad everything is and here you come along duplicating what the buffalo did 150 years ago. These are the kinds of leases I get excited about. The landowners could hardly wait for me to get started.

Things Are Starting to Click

Now I had 220 acres that made up one big square. It was going to make my paddock rotation a perfect circle. I spent a month of evenings getting the 70-acre complete perimeter fence installed. I was able to use my existing electric fence charger that was energizing the 150-acre farm to also power the 70-acre farm. Talk about a money saver. There was no meter to set, no extra monthly electric bill, no extra charger, or ground rod system, just hook on and go.

The fall seeding on the 150-acre farm failed due to the lack of rain. I planned to frost seed red clover on the entire farm in February. The landowners of the 150-acre farm decided they wanted to run some pressure water lines off the rural water and place hydrants behind all the ponds. I got approved for a DSP-3 (water establishment program through NRCS). I performed all the labor and the landowners paid for all the materials.

I ran 1200' of buried pressurized pipe on the blacktop pasture, and set two hydrants with rock water pads at each hydrant. That fed five 6-acre paddocks on the blacktop pasture. I set four hydrants behind all the ponds, with rock pads for the water tanks. I was set, there would be water in every paddock when the new ponds filled up. When it started raining, this farm was going to be like a sleeping tiger waking up.

I started working on the 70-acre water situation next and was able to incorporate the water system of the previous leased 150-acre farm into it, also supplying the 70-acre farm.

The front 35-acre part of the farm is fed by a hydrant from the 150-acre leased farm. I ran 1000 feet of above-ground 3/4" 160 psi black pipe under the single hot wire that divided the paddocks. It feeds seven separate paddocks. It cost $130 for 1000' of pipe, which feeds the entire front 35 acres.

The back 35 acres has access to the valve on the one-acre lake from the 150-acre farm. Previously this lake had been off limits because of the steep banks leading down to it from the 150-acre farm side. It would have caused an erosion ditch if cattle had been allowed to tromp down the center of the steep dam. The back 35 acres sits right below the lake. I just set a tank and put in paddocks. I could not believe how things were starting to click together.

Getting Paid to Clean Up Duff

I found 30 dry cows to custom graze on the 70 acres through the winter. The land had a tremendous fescue thatch built up on it. I strip grazed those 30 cows the entire winter on the 70 acres of fescue thatch. The cattle owner paid for three lbs of corn gluten every other day for protein. No hay was fed to anything, which made the cows eat everything before I moved the wire.

I made $900 grazing those cows that winter cleaning up the thatch, plus they fertilized the entire farm. They made a beautiful seed bed to no-till into for the coming spring seeding. The best part was I was paid to do it. If I would have hired a tractor and brush hog to mow the 70 acres, it would have cost a minimum of $20.00 per acre, or $1400 for the entire farm. There is absolutely no better way to clean up a lot of old duff then to strip graze dry cows and get paid for it. I would rather have the money going into my pocket than going out.

The landowners of the 70 acres were very interested in bringing up the fertility of the farm through a cost/share program. This was done the following spring. The landowners paid for all materials. I supplied the labor. All lime and fertilizer were put down while the ground was frozen. I spread no-tilled red clover on the bare ground that the cows had stripped off during the winter. It had a weedy bottom along the creek that the landowner wanted something done with. I put out 15 acres of Palaton reeds canary grass. I split this into 3 five-acre paddocks. They are fed water by the hydrant on the 150 acres also.

The Ultimate Surprise

That same winter the landowners of the 150 acres came by the house after deer season. He handed me a cured ham and frozen turkey. I was very surprised. He told me, "Greg this is just a small token of our appreciation for all the work you have done on our farm this year."

Their gratitude is hard to put into words, as well as the satisfaction that I received when he said that to me. But he wasn't through.

He reached in his pocket and gave me back my whole year's lease money in cash!

I refused to take it, but he wouldn't take no for an answer.

Then he saved the best for last. He went on to tell me that his wife and he had talked it over and they thought the ten-year lease was unfair.

A lump came in my throat initially, but then he finished the sentence.

He said," My wife and I have decided to give you a lifetime lease on our farm, not our lifetime, but yours."

I about fainted. You talk about excited, I was in total shock!

Drought continued through the fall, and none of the new seeding came up. So I frost seeded again in February, but we did not get any spring rain. I had eighty-four 400 lb steers stocked in April anticipating the spring rush of grass from the usual spring rains. I wanted to be conservative on stocking for the first year. I didn't want to run out of grass early and have to call the stocker owner to come and get them. Finally on June 8th, we started getting rain. It filled the ponds half full. Now each new pond site was a reliable water source.

The grass and clover took off like crazy. The steers could not keep up with all the grass. The whole farm looked like a lush salad bar. I grazed the group until September first. They gained 1.35 lbs per day. For the first year of stocker grazing I was tickled to death. I waited 40 days and got 85 heifers to custom graze that weighed 440 lbs for the fall grazing season.

When you work hard, give all you've got, and manage their property as if your livelihood depended on it, awesome results can occur.

I worked out a separate contract on the fall heifers for wintering them on the leased farm. I wanted to have some heavy calves in place to graze in April of the following year. These heavier calves were going to be my mowing machine to prevent the grass from getting mature. I wanted something that would hog off the grass.

I wintered the 85 heifers on the combined leased 220 acre farm, of which 150 acres was in grass. I fed corn gluten for protein every other day, and had big round bales set out in the individual paddocks for feeding.

All bales were unrolled with my truck and chain. I got some awesome fertilization from these rows of unrolled hay, and concentrated on the poorest areas of the farm.

By rotating calves through the paddocks in the winter, I only fed hay for three weeks.

At the end of the winter, I had fed half of a 1200 lb bale per 600 lb calf. The heifers had grown a lot of frame, were not overly fleshy, but were primed for the coming spring grass rush.

I lined up an additional ninety-five 400 lb heifers to be delivered May 1st. This is the time I always lose my grass to maturity. It seemed as if in 10 days time everything has a seed head on it. I had one awesome grazing machine now. I had them stocked at 175 head to five-acre paddocks. I could not believe the dramatic difference the additional 95 head made. They ate everything. Young weeds became forage. All sprouts were history. There was manure on every foot of ground.

Leasing idle land gives you an unfair advantage.

I talked with the landowners about fencing off all the woods on the property. I was losing a lot of manure piles back in the woods. It just killed me to see all those dollar bills (manure pile = 1 buck!) scattered throughout the trees. They would never grow any grass in the woods. The calves in the fall always hogged down on the acorns. I had some sleepless nights worrying about them getting acorn poisoning.

The landowners brought their three sons up and we fenced off all the timber. Their sons are starting to learn about the importance of MiG on the farm as well. I explained to them that they could expect to see more deer now that the cattle were not in the woods.

I also volunteered to help them with a timber stand improvement program. I offered to cut all the undesirable trees to open up the canopy for the money trees.

The landowners wanted to build two more ponds on their farm to help me with a more reliable water source in dry times. Three of the remaining paddocks still relied on a creek for water. They offered to pay for the two ponds but let me decide where I wanted them.

Taking a vacation for these particular landowners meant

coming up to their farm, cutting, burning brush, snipping, painting thorn sprouts, helping me build fence, you name it and they were eager to do it. Anything they wanted to do on the farm, I did my best to make it happen. We have built a very strong relationship over the years.

The bottom line is, when you work hard, give all you've got, and manage their property as if your livelihood depended on it, awesome results like this can happen.

Chapter 8
Writing the Lease Proposal

Before you write the lease proposal, here is the list of information that you should have researched and the list of items that you should have already explained to the landowner:

1. A legal description of the property.
2. The number of grazeable acres calculated accurately from the local extension office.
3. The number of years the landowner is interested in leasing.
4. You have explained in detail to the landowner the advantages of MiG (extremely important).
5. You have already shown the landowner a working MiG farm, preferably one of your own.
6. You have explained in detail to the landowner the work and management that will be required to implement MiG on their farm.
7. Calculations of the potential payback of the leased property.
8. Calculations of the cost of implementing a grazing system.
9. Know all the landowners' concerns about the property.

Now that we have covered all the above items concerning the property, let's write the contract proposal. Every contract that you write will look a little different, but you want to cover the landowner's concerns and your own 100%. Remember this is your first proposal. It is not set in concrete. Look at this as a starting point, then the landowner and you can fine tune any additions or retract anything you want.

John Doe Land Lease Contract Proposal

I, Greg Judy, agree to lease from John Doe, his farm, which consists of 150 acres, located in Howard County, Missouri for a period of 6 years. The legal description of the farm is Bourbon Township, Section 10, Range 7. The lease will start the day the cattle are placed on the farm. At the end of the 10 year lease contract, Greg Judy will have first chance at a new lease, if John Doe is satisfied with the way the farm is being managed. Greg Judy agrees to have a liability policy in effect by the cattle owner before the cattle are placed on the farm. Greg Judy is responsible for any bodily injury that would occur to himself or others under his employment.

The lease payment is due every six months from the date the cattle are placed on the farm. Leasee agrees to install a Management-intensive Grazing system on John Doe's farm. Leasee agrees to do all the labor for installing additional electric fence and maintaining the existing fence during the term of the lease. All wire, posts, and water tanks are to be provided by leasee and shall remain his property at the end of the lease. Landowner agrees to give leasee 90 days to remove all fence, posts and water tanks at the end of the lease.

All ponds will be fenced off with single hi-tensile electric wire to exclude cattle. Leasee will run siphon hose over the dam, hooked to the stock tank for watering livestock. Leasee agrees to provide all labor to clear all brush and paint all tree stumps where new paddock division wires will be. Leasee will install single hi-tensile electric wire paddock divisions, where appropriate, to allow for grass rest periods between grazing rotations. Leasee will frost seed red clover in February each year to maintain a vigorous stand of legumes in pastures.

Leasee agrees to control all trespassing that may occur. Leasee will not fasten any fence to any tree unless the landowner is in agreement. Leasee has 100% control over all grazing and management decisions.

Year 1 = Will repair all perimeter fence and remove brush where needed. Will fence off all ponds, add siphon hose hooked to stock tank. Will run single wire paddock divisions where appropriate. Will use cows to clean up old thatch the first year. Will take soil tests on pastures to monitor health of soil.

Year 2 = Will frost seed red clover in February on paddocks. Will control brush invasion on pastures by mob density grazing. Will fence off all timber areas to exclude cattle. Will rotate calves through paddocks to prevent overgrazing.

Year 3 = Will assist landowner in any water or pasture improvements. Will continue to carefully manage the grazing, which will improve the soil fertility, earthworm populations, increased forage species, along with increased wildlife populations on the farm. Frost seed clover in February.

Years 4-6 = Will continue to carefully manage grazing to maximize farm potential. Will control unwanted brush invasion. Seed clover as needed to maintain healthy stand.

Lease Payment Schedule:

Years 1-6 = (You should have a good idea of how much you can offer, yet still leave yourself a good profit. Just repeat the calculations that I went through in chapters 6 and 7 on setting up a MiG system).

Leasee Signature Date:

Landowner Signature Date:

Comment:

Please voice any questions, concerns or comments that you may have with the above contract proposal.

That is about all that I cover on an initial contract proposal. You may have to negotiate back and forth on the lease price or other topics. Do not let your emotions with wanting the property so badly outweigh your common sense. You have a rough idea what the costs are of implementing the MiG system. Make sure the landowner knows that you are taking all the risks as far as costs go. You're paying for all MiG inputs in the above contract. If the landowner wants his place to look like the MiG place that you showed him, he should realize right at the start that it is going to cost you a considerable investment to do so.

Treat landowners like you would like to be treated.

Remember we are not talking about leasing some land, throwing some cows on it and checking them every couple of weeks. We are building paddocks, preserving the water, getting awesome manure distribution, which builds the soil fertility, stimulating the soil bank that has been in hibernation for years, introducing legumes, allowing the earthworm population to explode, encouraging the explosion of grass species, seeing increased wildlife numbers and species, controlling brush by mob grazing, and catching more rain water which cuts down on erosion. These are just to name a few improvements.

A landowner would have to pay a fortune to hire all of the above done on an annual basis to keep his place looking nice year in and year out. To get the two to three dollars per acre a year lease, you have to make the landowner aware of all the additional inputs that he is getting as a direct result of your labor and management.

If a landowner doesn't seem interested in improving his land after I have explained and showed him a working MiG farm, then I thank him for his time and go elsewhere.

If a landowner is just inexperienced about grazing in

general, I can explain MiG to them. After it is explained to them, sometimes they get excited and want to be part of it. Folks, MiG is plain natural. It is just duplicating what the large herds of buffalo did for centuries. It does not take an educated person to realize what monstrous effects a large herd of animals do to a five-acre paddock in one day. Go walk it, you had better watch where you place your feet! Then walk the same paddock 30 days later. All manure piles are basically gone, complements of the dung beetles and earthworms.

You have to be convinced yourself that MiG works before you can convince a landowner that it does.

You need to become a MiG salesperson to explain the complete picture in the mind of an uneducated landowner. It is easy to sell something that is new, natural, in-sync with nature, stimulates wildlife and last but not least, *works.* You have to be convinced yourself that MiG works before you can convince a landowner that it does.

If you do not have any MiG experience, read up, go to field walks, attend workshops, do whatever it takes to get yourself educated in the field of MiG. Once you are armed with the knowledge of how MiG works, then tackle writing your first lease.

Start with a small section of leased land, don't overwhelm yourself at the start. Learning MiG with cows and making mistakes, is the fastest way to learn. Then move up to stockers. They are not as forgiving as a cow. If your grass gets mature the cows will get by eating it, they won't like it, but they will eat it. Stockers will lose weight if you force them to eat highly mature, unpalatable forage. If you're getting paid by pounds of gain, this is not the way to make money.

The best part of writing a contract is you have no risk. It's like asking a girl to go on a date. You don't know if she will go unless you ask. She will either say yes or no. It is exactly the same with writing a lease. The only thing you are losing is the time invested in writing the lease proposal. Don't be disappointed if you fail to obtain the lease. Chalk it up as valuable experience; go on to the next land prospect.

I had my eye on a piece of ground that was across a gravel road from one of my leased farms. There were 80 acres, 70 of which were open. The landowner lived in town. I hardly ever saw her on the property. I met her boyfriend who lived with her in town at the farm one day. He was about 45, and a permanently unemployed alcoholic, I found out later. He mentioned that the farm I had leased was really looking good. He made the comment that he would like to possibly do something with his girlfriend's property.

I agreed to walk the property, then write a proposed land lease for them both to read over. I wrote the contract, and gave it to him the next time I saw him at the farm. When I handed him the contract he never even bothered to read it. He threw it behind his truck seat and asked me if I wanted a sip of whiskey. Later I found out he never showed or read the contract to his girlfriend (the landowner). I found out later the farm is his party pad, where all his friends can come out and drink. If I were to start showing improvements on her property, his girl friend would probably kick him out. He was telling her that he was going out to her farm to work, but no work ever got done. This is one lease I'm happy I didn't get.

Writing a contract proposal has no risk. The only thing you've invested is time.

If you smell whiskey on the landowner's breath before 7:00 p.m., walk away. No matter how hard you try to please someone who is under the influence of alcohol, you will never satisfy them. No matter how much land an alcoholic owns, leave him alone. I knew a fellow grazier who leased a huge grass farm that was owned by an alcoholic. The landowner had inherited the whole farm from his parents. He would brag about his method of rounding up his cows. He would get in his 4-wheel drive pickup, load up his shotgun with birdshot and go round up the cows. When the cows saw anyone coming at 100 yards they took off charging into the brush. I can't say as I blame them. His entire farm was severely grown up in brush and had no fence. The place was just a wreck. The farm had not been brush hogged for twenty-five years. The alcoholic landowner agreed to supply the grazier with the fencing materials if he would do all the labor. The grazier fenced the entire farm with five strands of barbed wire, cut brush, really gave the farm a face-lift.

Be honest, work hard, and land may be offered to you.

After several years, the alcoholic decided that he was going to sell his farm. The grazier had to get out. One month after the alcoholic sold his entire farm, he died of a heart attack while he was purchasing his morning bottle at the local liquor store. Here was a fellow who had this beautiful 700-acre grass farm handed to him, stocked with 100 cows. He had a good barn, a corral with weigh scales, a hay barn, everything in place. He never had to work for any of it. Everything was given to him. That was his doom.

I believe if you don't work for it, you have no appreciation for the work and sweat that it took to get it. The things in life that I've accomplished that caused the most sweat are the things that I appreciate the most.

Alcoholism is a disease. Alcoholics want you to feel as miserable as they do. You don't need this kind of torment in your grazing operation. There is nothing wrong with drinking a cold beer, but 12 hours everyday of your life, I think, is excessive. I know several landowners who inherited their parents' farm and ended up drinking the whole estate away. They never did anything with the land except let it grow up in brush, then sold it for more whiskey money. That's a heck of a way to treat your parents' farm after they worked their whole life paying for it in order for you to have something when they are gone. Life is too short. There's no need to hurry it along by drinking yourself to death.

Writing a lease proposal is a lot less stressful than going through all the paperwork involved in buying land. I remember my first land purchase. A little voice in the back of my mind kept nagging at me saying, "Everything had better go right or you will lose your house, the farm and everything you put in it." I was wondering what would happen if I lost my job in town, or got sick. You name it, I worried about it. I remember walking out of the land closing office, feeling like I had signed my life away.

What a wonderful feeling it is to 100% manage a piece of ground and not have to worry about how you're going to pay for it. No land title company, no banks, no real-estate sales person, no closing costs, no taxes, no down payment. Leasing land is just a great no-stress way to get in the grazing business for anybody, period. Grazing livestock is a great way of life, but eliminating the stress of land ownership makes grazing just that much more enjoyable.

What a wonderful feeling to manage a piece of ground and not worry about how to pay for it.

When the lease runs out, if you have done a good job of managing the land, there is a very good chance that the landowner will renew the lease with you. If he doesn't, go find some more idle land. Who knows, you may find a lifetime lease. Lifetime leases are out there. I proved that. The beauty of a lifetime lease is you can make management decisions like you owned the farm. Everything you put into the property, you can reap the rewards. It's yours to manage for life.

When writing a lease proposal on a piece of land that has a lot of grass and no water, get an estimate of what it would take to put in a pond. Write in the contract that you will pay for the pond if you can deduct it off the land lease payment. Put a pipe and valve in the pond during construction and you've got water. The pond adds tremendous value to the land. Everyone wants a pond on their land for fishing or other recreational activities. I know of a farm that sold for triple the local land price, all because it had a pretty pond right in the middle of the property that offered a beautiful view. People like water; it is essential for life. Offer to stock the pond for the landowner. You can usually get the fish free from your local conservation department.

When writing the lease, you need to show that you're enthusiastic about the prospects of the land. A comment like, "With some good grazing and management practices, I can turn your land into a very valuable investment." Enthusiasm is contagious.

Do not make a list of promises that you can not keep. Be 100% honest with everything that you put in writing on the contract. Be 100% honest with the landowner on every little detail. If you make a mistake, own up to it. The landowner may get upset, but he will respect you for being honest with him. Treat the landowner like you would like to be treated. Be patient with them. Remember, Rome wasn't built in one day. A steady slow pace to building your grazing business will result in reaping the rewards for years to come.

Chapter 9
Developing Good Water

In this chapter, we will go over some cost effective ways that I have used to develop reliable water sites in a MiG system that will not break down from daily use. Water is always the limiting factor when setting up a grazing system. A good water system is the single most important factor in developing a good grazing system. You can have the best grass, awesome new grass species, beautiful fences, a sturdy corral, good road access, utilities, the list goes on and on. The bottom line is, if you do not have a good reliable water source, you're dead before you get started.

A MiG system puts a lot of pressure on the watering site. Remember, you may have several hundred calves coming in to drink within a relative short time period. Take 200 animals x 4 hoofs = 800 hoof prints around the water tank. This cycle is likely repeated three to four times each day. Now add a three-inch rain in two hours to the equation. You end up with nothing but soupy mud in a 25-foot circle around the watering site.

If you force a calf to wade through mud just to get a drink, then the calf is going to limit the number of times that they drink. The mud is also a good place to harbor and transmit diseases. It makes me sick to see calves standing in mud up to their bellies, just trying to get a drink. This will cut your daily weight gains back also. The more water they drink, the cooler their body temperature, the more grass they will eat. Do your calves a favor, at the same time you will be putting money in your pocket. Form a solid pad out of rock before you unload any calves. It is a small investment for such a large payback.

If you don't have a good reliable water source, you're dead before you get started.

I use geo-textile fabric with a load of 1-1/2" rock dumped on top of it, which is very economical. It makes a good smooth base. The more they walk on it, the harder it gets. The geo-textile fabric is used on construction sites a lot. The contractors lay down a layer of it on spongy new ground, and cover it with rock. Large concrete trucks can back right up to the building site without sinking to China. It is made from polyethylene material which is very tough, and hard to tear or rip.

The geo-textile physically separates the soil and rock preventing the rock from settling into the soil beneath it. You need a minimum of 5" of rock on top of the textile. I used waste rock, which is 1-1/2" rock all the way down to lime dust, but it doesn't stay in place very well. Now I strictly use 1-1/2" clean rock. The geo-textile comes in 12' wide rolls, 400' long. It costs around 70 cents a foot. I lay a 20-foot piece of textile on the ground at my selected water site. I weight down the edges of fabric with brick, rock or sand. This weight prevents the fabric from moving when you have the gravel truck dump the rock in the center of the fabric.

The very first pad I made, the fabric had no weight on it. When the truck backed up on the fabric, it moved around a lot. From that time on, I always weighted down the edges and the center. I never had any more problems with fabric moving. I order an eight-ton load of rock for my 20'x12' pads. Counting rock and geo-textile, the entire pad costs $80. This same pad made from concrete would run around $600.

Another cost effective water pad is to look around construction sites where they have torn up sections of concrete. Some of the sections are in 4-7' square slabs. The best part is they are free because the contractor has to get rid of them. If you take them off their hands, they don't have to pay a dozer operator to bury them. You can push them together at the water site with a front-end loader. It makes a very durable water pad. These concrete sections are also very handy in loading areas or for setting feeders on.

You need a good clean dependable water source. I can not over emphasize this enough. Keep the cattle out of their water source. You wouldn't want to drink water that had been manured, urinated, waded, and lounged in. Livestock should not be made to either.

If given a choice of clean cold tank water or dirty contaminated water, livestock will walk long distances to drink the clean cold water. Remember, the younger the livestock, the better quality your water needs to be. An old cow can drink some pretty putrid water and still perform, although I don't like it. But with stockers, you're asking for trouble if you force them to drink putrid water. You will pay for it by increased disease and decreased gains.

You need a good, clean, dependable water source.

I love to watch a young, grass fat steer come into a water tank that is setting on a rock pad. He will draw his mouth down into the nice clean cold water.You can watch his throat just swallow big gulps. The steer almost has a smile on his face when he walks away. I know I do.

I have noticed that the calves that are on piped rural water will be up grazing in the heat of the day, while the calves that are on pond water at other farms are lying in the shade. The only theory that I can come up with is the rural piped water is always very cold and the calves just love it. They drink a lot of it, which helps keep their body temperature down in the severe heat.

First look for the water resources already on the land.

Pressure from the water line allows me to have the tanks centrally located in the paddocks. Several calves are always dropping by for a nice cool sip of water. I have never regretted spending money on a good water system. It will pay you back many times your investment.

There are some cost-effective ways of securing water for grazing. First look at what resources are at hand. Are there any creeks, ponds, springs or old wells on the property?

Creeks make a good water source as long as you can limit access to them. If the cattle have full access to the creek, the water will not be worth drinking. They love to lounge in a creek. That's where I would be if it gets hot. There are usually lots of trees for shade, the creek water is cool, and it gives them a place to escape the hot temperatures. By July, the water has turned to a black mire muck due to the lack of fresh rain water flushing out the stale putrid water. Now you've got problems. Calves get sick, lose weight, have pink eye problems, the list goes on and on.

If you use a creek, pick out a good solid gravel bar and run a hot tape around the area. Cattle can get a drink, but are not allowed to foul up the entire creek. Before you count on the creek for a summer water supply, ask neighbors around the area if the creek ever dries up. It is a sick feeling to run out of water and still have grass to graze.

Another word of caution about creeks. You hear more and more legislation talking of livestock operations that connect to creeks. Government agencies are going to start monitoring the creek water that comes from those particular operations. There is talk that they will be able to fine you or completely shut you down if there are violations. Lot of rumors are floating around, but I predict something along these lines are in our immediate future.

Our conservation department has cost/share programs that pay the majority of the cost of developing an alternate watering source for cattle, if you agree to fence the cattle out of the creek. This is a wonderful program for landowners. It allows them to build a reliable clean water source on their property. It also removes any future problems with cattle manuring in the creek.

I know a lot of people who get almost violent when you start talking about keeping cattle out of the creeks on their property. Their response is, "By god my grand-pa, dad, and I have watered cattle out of that creek for 70 years. I will be darned if anybody is going to tell me that I can't."

We can help ourselves though. Just be responsible and give your cattle limited access to creeks. I use creeks on four of my farms. With limited access I get by fine.

I will take a pond over a creek any day. One huge advantage of ponds over creeks is not worrying about the water gaps washing out. When you have several farms that are using creeks, it is a mad dash from one to the next rebuilding gaps after each major rain. You just pray the cattle are still in.

If you won't drink it, don't expect your animals to drink it.

Ponds are easy to hook onto, and make very reliable

free water sources. A very functional pond system would be to run a single strand hot wire around the pond. Run 3/4" polyethylene siphon pipe over the dam hooked to a seven-dollar water tank float. It doesn't get much cheaper than that.

You must keep the cattle out of the pond or you have wasted a valuable watering site. Cattle are like small bulldozers when they enter a pond. Every time they slide into one they make the water level shallower. After several years, you have a worthless mud pit. Cattle are almost as hard on ponds as hogs. The water is always muddy and brackish, not ideal stocker water. It amazes me to see landowners hire a dozer to push out all the mud from their pond, then turn right around and let their cattle have access to it again. Talk about wasting good money. They could have easily put a water pipe in the dam when they had the mud pushed out.

Ponds don't have water meters on them. It is free water. When watering large herds of stockers, this is a nice savings. Build a pond if the landowner agrees you can deduct the cost from the lease. If you build a pond, try and place it where you can use gravity flow to feed other paddocks located below it. Are there any filled-in ponds that could be cleaned out in several hours?

I leased a farm that had an old pond that had the dam cut on it several years earlier. This pond had been a former pig pond. The last owner had allowed pigs in it until it wouldn't hold water, just mud. The old pig pond was located in a perfect little draw that sat close to the top of a hill. I could envision it feeding at least three paddocks if I had it cleaned out. I got a hold of my dozer operator and he made a beautiful pond out of it in four hours. I laid a 1.5" schedule 40

You must keep the cattle out of the pond or you have wasted a valuable water resource.

PVC pipe in the dam's bottom. This was hooked to a hydrant, then trenched down the hill and fed two other paddocks as well.

Make the pond deep. You do not need a lot of surface area. You're not going to be skiing on it, just watering livestock from it. Most of the water that comes out of a pond is from evaporation, not livestock consumption use. The shallower the pond is, the quicker it will evaporate in a drought. Shallow water heats up faster and accelerates evaporation, so make 'em deep.

Make your ponds deep.

Another advantage of deep ponds is the water coming off the bottom is colder. Stockers will drink more of it, which makes them gain more because they eat more.

I used to think that every pond that I built had to be one to two acres, or it wasn't a pond. One day an old timer asked me, "Son why do you want to cover all that grass up with water? Cows will never be able to graze that area again."

That one comment changed my whole mindset about ponds for livestock. I had never thought of it that way. Grass gain is what makes money, not two to three acres of water that just covered up your grass. The bigger you make the pond, the bigger your pockets better be to pay for it also.

Walk the farm and look for slight draws that are centrally located in the pasture. Beware of building ponds in any small branch. A friend of mine dammed up a tiny innocent looking branch. When it rained the fish could actually swim up the branch out of the pond. The dam eventually washed out. He had an 80 acre water shed dammed up.

When selecting a pond site, look up above the pond area and visualize how much area is going to drain into the pond. The area should be at least twice the size of the potential pond site. For a 1/4 acre livestock pond, you would not want anymore than three to four acres draining into it. Remember, once the pond is full, every drop of water from the watershed that runs into the pond has to flow through your spillway.

Each pond can easily handle four paddocks if it's located correctly. Express your concern to the dozer operator that you want the pond deep, with minimum surface area and you want it completed in 6-7 hours. I have had a couple of nice ones built in four hours. Both had a nice short draw to work in.

Each pond can handle four paddocks if it's located correctly.

In my area, 6-7 hours of dozer work will get a pond 7-10 feet deep, depending on the location. Most of my 4-7 hour ponds are less than 150 feet wide, but the edges drop off like a bluff. Ask the dozer operator to divert the spillway away from the back of the dam. You do not want the water to drop right off behind the dam. If water goes straight down the backside of the dam, it will start cutting a ravine. In time the ravine will take out the dam.

If you do not have any experience with dozer operators, ask to see some of their work. Talk with the owners of some of the ponds that the operator has built and see if they were satisfied with his work. It is a lot better to get a good operator on the job site at the start, than to tell him to get lost after he has messed up the pond and has racked up multiple hours. Have the dozer owner give you a written bid and make him stick to it.

From my experience, the bigger the dozer is, the more dirt he can move, which makes it the most economical. There is a huge difference in dozers also. I had a contractor give me a bid on a pond site, which I accepted because I thought it was fair. When they brought the dozer out, it would fall on it's face every time the operator tried to push his blade in the dirt. He worked on the pond for a week and still did not have the dam close to being finished. The contractor finally brought out a second dozer and they teamed up on it until it was finished.

Luckily I had gotten a firm written bid from the contractor, otherwise the pond would have cost me a fortune. I guarantee the contractor did not come out too good on that pond.

There is a huge difference between dozer operators as well. I had one dozer operator who had a beer open in the morning when he pulled up. He drank two cases each day while he dozed on the bid job. He was a good operator, but if you tried to watch him build the pond, he would just pull his dozer up in the shade, pop a beer and shoot the breeze. Needless to say, I never had him back.

I found a good honest hard working dozer operator. I use him exclusively on every pond that I build. There have been several instances in which he has changed my mind on where to put the pond site. He will reply, "Greg, if you will let me move down the hill to that tree, I can get you another four foot of water for the same money." He is very accurate in his time estimates on building the ponds also.

The operator should be able to fill the 12-foot blade and have dirt rolling over the top of the blade while he is pushing it toward the dam. That is when you know you're getting your money's worth. I do not like hi-lifts for building ponds. I am convinced that a dozer with a 12-foot blade will get a pond finished faster and the dirt will be packed better when it's finished.

Put a minimum of 1.5" plastic schedule 40 pipe in the bottom of the dam. I can not stress this enough. Use schedule 40 PVC pipe. It is strong enough that with some dirt over it, the bulldozer will not collapse it when he is running the dozer back and forth over it. I have seen ponds that were built where the owner had put in 3/4" black polyethylene pipe. The pipe is collapsed or

Put a minimum of 1.5" plastic schedule 40 pipe in the bottom of the dam.

broken off, completely worthless. If you're going to go through the exercise of building a pond, put a good plastic pipe in the bottom of the dam.

The day the dozer shows up to start on the pond, I already have a minimum of 100 feet of schedule 40-1.5 inch PVC pipe ends primed and glued together. On one end I have a 90-degree plastic elbow with a six-foot piece of 1.5" pipe glued to it. This six-foot piece has a plastic cap glued on top to prevent fish, leaves, sticks, etc. from clogging it. I then take a cordless drill and drill 50 1/4" holes in the top section of the six-foot riser pipe. This gives you very good pressure at the water tank, while preventing any clogging of the pipe.

POND DETAIL

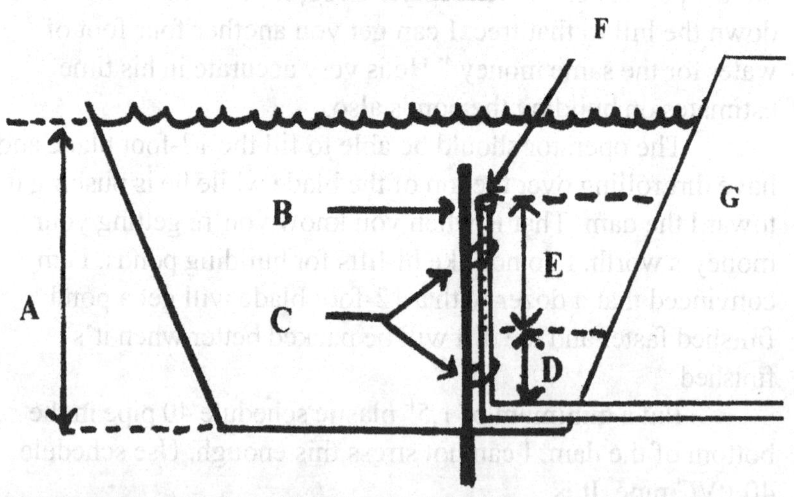

A - 8' minimum water depth
B - 1" fiberglass post 8' long
C - 2 - ½" Nylon straps
D - 2'
E - 3' (drill 50 - 1/4" holes into top 3' section of pipe)
F - 1-1/2" PVC schedule 40 cap
G - Pond dam

When you finish drilling the holes, make sure that you remove all PVC drillings before gluing connections together. I had a water pipe that wouldn't flow very well after the pond filled up. After running a snake through it, I pushed out a wad of PVC pieces that were created from drilling the holes. Make sure you do not drill any holes in the bottom two-foot section of riser pipe. This area is needed for soil settling during fill up.

The dozer operator never gets off the dozer. He digs a trench in the bottom of the dam. I lay the pipe and fasten it securely out in the pond to the fiberglass stake. I try to get the riser positioned in what will be the deepest section of the pond. The dozer operator can go on pushing dirt while I finish the plumbing on the back of the dam. On the back of the pond, I glue in a plastic shutoff ball valve. I place an 8-10" diameter piece of plastic tile, 3 feet long around the valve and cover the sides with dirt. Directly down from the shutoff valve, I place a frost free hydrant. The dozer operator backfills everything while he is dressing down the back of the dam. The beauty with this method, all dirt work and plumbing are complete when he leaves. There's no backhoe or trencher work to hook up anything. It is all done.

List Of Materials Needed for Pond Building

1. Schedule 40 1-1/2" PVC pipe. Estimate length needed then add an additional 20 feet.
2. 1 schedule 40 1-1/2" cap.
3. 2 schedule 40 1-1/2" elbows.
4. 1-1/2" brass shutoff valve.
5. 6" to 12" diameter pipe, 32" long.
6. Hydrant.
7. 8" wooden post, 8' long.
8. 1"x 8' fiberglass post, 2 nylon rope pieces (2' long).
9. 7 ton rock.
10. 12'x18' geotextile fabric (Optional).

Installing a Pond Water System

(See Pond Sketch on page 97 for overall view.)

1. Cut a 5' length of schedule 40 PVC pipe for riser pipe.
2. Drill 50 1/4" holes on the top 3' section of the pipe.
3. De-burr the inside of pipe where the 1/4" drilled holes are. These PVC shavings on the inside of the pipe will clog up your hydrant and stock tank float. A piece of sandpaper taped to a stick works great for removing PVC shavings.
4. Apply a liberal coat of PVC primer to all fittings.
5. Glue sections of pipe together and lay them in the bottom of the freshly prepared dam site. Allow the end of the pipe to extend out to the center of the pond.

Pond Sketch Key

A - 8' minimum water depth
B - 1" fiberglass post 8' long
C - Two ½" Nylon straps
D - 2'
E - 3' (drill 50 - 1/4" holes into top 3' section of pipe)
F - 1-1/2" PVC schedule 40 cap
G - Pond dam
H - 1-1/2" Brass shutoff valve
I - 1-1/2" schedule 40 PVC pipe
J - Shut off valve casing, 6" to 12" plastic pipe, 32" minimum length
K - 8" diameter wood post, 8' long
L - Water hydrant stepped to post
M - Livestock water tank
N - Geotextile fabric, base is 12' x 18'
O - 4-6" layer of 1-1/2" rock over Geotextile
P - 1-1/2" pipe
Q - 2' between Shut-off valve casing and 8' wood post
R - Side view of Shut-off valve casing

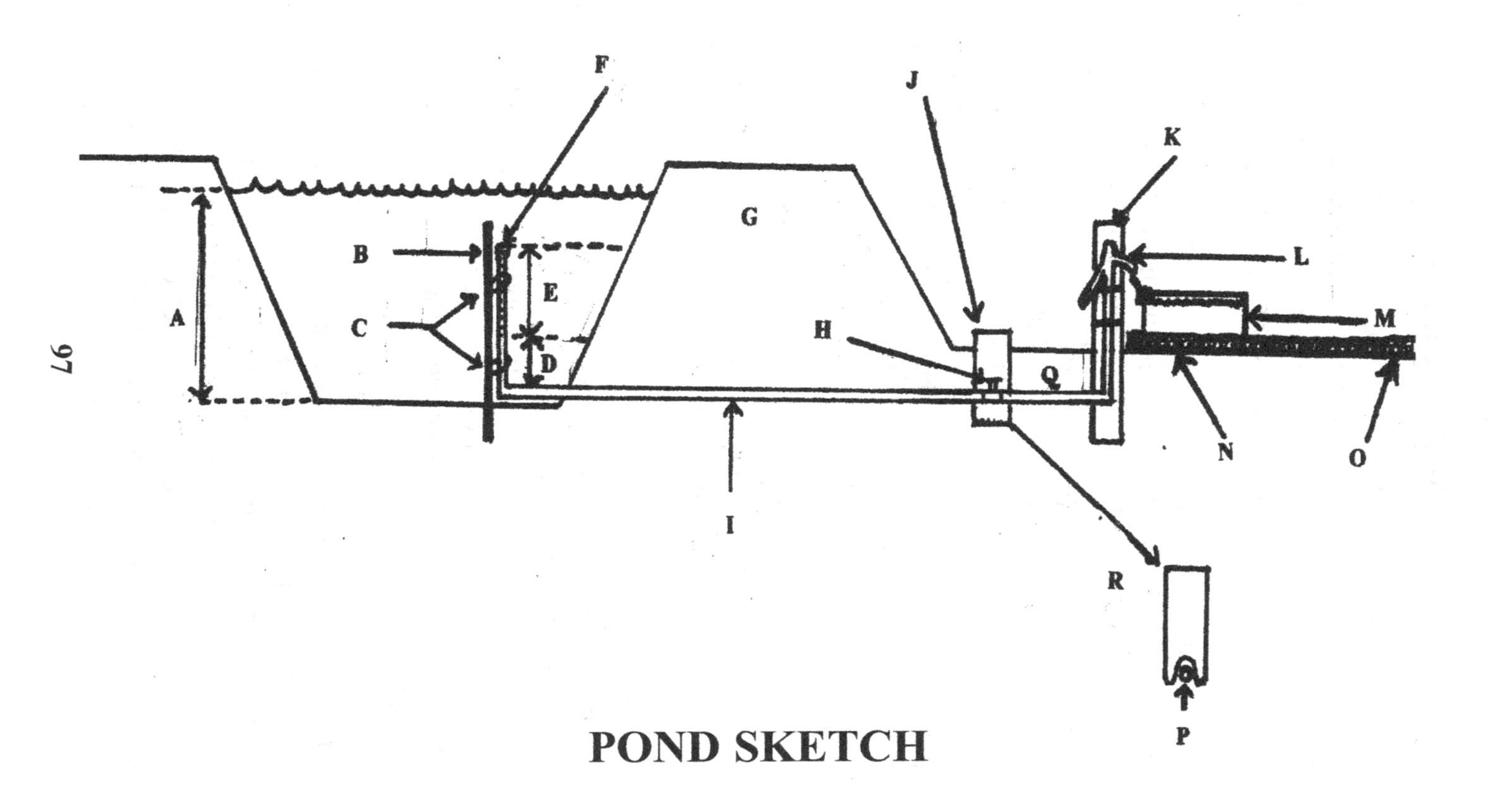

POND SKETCH

6. Tape the end of the pipe to keep dirt out. Drive a temporary post at the end of the pipe and tie a white ribbon to the post. This gives the dozer operator a very good visual marker to keep him from burying the end of the water pipe during the construction of the dam.
7. Glue on a shut off valve.
8. Glue on a 2' piece of 1-1/2" pipe that hooks to a hydrant.
9. Cut a V in each side of the pipe casing that allows the pipe to enter and exit where the shutoff valve is protected.
10. Place the pipe casing over the shutoff valve, then drive a stack on each side to hold it in place until backfilled with dirt.
11. Pour two five-gallon buckets of rock at each side of the pipe casing that surrounds the shutoff valve. This allows ground water to drain away from the shutoff valve.
12. Dig a hole, set and tamp an 8" diameter post 8' long at the hydrant location.
13. Lightly wire the hydrant to the post.
14. Glue on the final fitting that hooks to the hydrant. Tighten a wire to secure the hydrant to the post.
15. Pour two five-gallon buckets of rock at the base of the hydrant for the drain field.
16. When the dozer operator finishes the dam, have him backfill around the shut off valve and hydrant.
17. Have him scrape off a 20' x 20' drinking tank pad in front of the hydrant. Have it slope away from the hydrant slightly.
18. Remove the tape from the end of the pipe that is located in the center of the pond.
19. Glue on an elbow, then a five-foot riser pipe, then the cap.
20. Drive the 8' fiberglass rod next to the riser pipe and secure it with the two pieces of nylon rope.
21. Lay down geotextile fabric on the drinking pad. Secure the corners with bricks until the rock is placed on the fabric.
22. Place a 4" to 6" layer of 1-1/2" rock over the fabric.
23. Once the pond is finished, run a single strand of hot wire around the entire pond and dam.

I spread fescue seed on the entire dam and spillway area. Then I put some old hay on the dam and spillway area. The hay holds the moisture for the young grass seedlings. I will fertilize the spillway area with rotted manure or triple 13 to ensure a vigorous start of fescue. No other grass I know of builds a heavier and stronger sod then fescue. A solid stand of fescue in the spillway is absolutely the best way to prevent a gully from forming. It is a lot easier to prevent a gully from starting, than it is to stop one after it has a big ugly cut in your spillway.

Allow room between the hot wire and pond for a person to comfortably walk and fish around the pond.

I used to car pool with a guy who owned a pond that had spillway problems. Every time it rained, he about killed us driving like a maniac to get home and stuff something in his spillway. He had a gulley in his spillway that you could drop a dump truck in.

One pond that I built, I made the mistake of hauling in big rocks, called"knurls." These were placed in the spillway area to prevent erosion when the water came through the spillway. The rocks did a very poor job of stopping the erosion action. The water just went around or cut under the rock leaving big cutout holes. I hauled in rotted manure, and heavily seeded it with fescue. That cured it. A thick root system holds soil better than anything else that I know of when water is running over it.

Once the pond is finished, I immediately get a single strand of hot wire run around the entire pond and dam. This prevents any cattle from tromping big holes in the freshly exposed dirt. I leave enough room with the hot wire that a person can comfortably walk and fish around the pond.

If the landowner is interested in having the pond stocked

with fish, then have the dozer operater leave some tree stumps in the shallower parts of the pond. If no stumps are available, make some homemade fish attractants from wooden pallets. Drive some steel posts where they form a square, then tie the oak pallets to them to form a square. These should be placed in water four to six feet deep. Fish love them for cover. It gives bait-fish a place to hide from the predator fish.

One well-placed hydrant can supply a lot of paddocks with good clean water.

State fisheries did a study where they built two ponds of the identical shape and size. One pond was left bare. Absolutely no cover was left in the pond after completion. The other pond had mature dead cedar trees anchored every thirty feet around the pond in 4-6' of water. Both ponds were stocked at the same time with the exact number and species of fish. Weight gain was measured periodically on the fish from both ponds. The fish gain in the pond with cover was triple over what their cousins' weight gain was in the pond with no cover. The pond with all the cover allowed the bait-fish a place to escape, where they could breed and populate like crazy. This gave the predator fish an all-they-could-eat smorgasbord. Not much effort or cost was involved in producing a magnificent fishing pond. Landowners are tickled to death with a healthy fishing pond also.

In some instances, the pond that is present on the farm may not be suited to set a water tank at the back of the pond. For instance, the pond dam may be really close to the property line, which could cause problems between the neighbors.

Another bad situation is if all the ground leading to the back of the pond is really steep, which with time will cause erosion ditches to start. The area behind the pond may be solid timber, which would cause the cattle to lounge in the shade, manure and destroy the watering site.

In this instance, use limited water access skirts that are built into the sides of the pond. Pick a fairly level area leading into the pond to locate the skirt. Haul in a load of 3" rock and dump it at the edge of the pond. Use a tractor front-end loader and spread it out into the pond 4-6'. The water should be 3-6' deep in this area. Drive a fiberglass rod in the water at each corner of your planned water skirt, 6' from the bank. Then run white poly tape around the whole skirted watering area. The 3" rock makes it uncomfortable for the cattle to walk on. This prevents lounging in the pond and approach area. A 20' wide skirt will easily handle 150 head. This watering point can feed three to four paddocks in a fan shape. It makes a nice economical watering site. The most important part is, you are preserving the pond and providing good clean healthy drinking water for your stockers.

Some ponds have severe infestations of moss, lilies, cattails, etc. I have found that three to five grass carp can completely clean up a pond in two years. They are tremendous foragers and will not reproduce or muddy up the pond. They are top feeders. Anything green is dinner to them. They can eat their body weight in forage per day. An extra treat comes when they finish cleaning up the pond, they are absolutely delicious to eat. They do not taste like the traditional carp. It is all white flaky delicious meat.

A word of caution when you try to catch them; have a really stout pole, hook and line. I've never had a fish on a hook that fought as hard as these dudes. They are tremendous fighters, but will bite on canned corn. Just chum an area with a can a day for several days. Then sneak back with your fishing pole and fasten on a couple kernals of canned corn.

Piped pressurized water is an awesome MiG tool.

Throw your line out in the area that you have been chumming and hold on to your pole.

Check to see if there is rural water, and investigate what it would cost to set a water meter. With one well-placed hydrant you can feed a lot of paddocks with good clean water. Once you have a hydrant, you can run hundreds of feet of above ground piping to strategic locations. This will encourage more uniform grazing and manure distribution over the paddocks.

Fence off springs to keep the cattle out.

Piped pressurized water is an awesome tool in MiG. It gives you so much flexibility in your grazing operation. You can pick out the poorest spot in your pasture, and set a tank at that location. After several grazing rotations, the whole area will be smothered in manure piles.

How many times have you wished that you could spread all that manure that is around the watering site onto the poor areas of your pastures? Piped pressurized water allows you to do that. It is an awesome fertility tool.

I get by with one small plastic tank for one entire farm, which eliminates the need to buy a tank for every paddock. I can run 100 seven-hundred-pound steers on one 40-gallon quick flow tank. With pressurized water, it refills as fast as they can drink it. It allows five to six steers to drink at a time. The small plastic tank is equipped with a Jobe Megaflow valve. I love it. The only moving part of the whole valve that the cattle have access to is the yellow bobber float. It dangles from a nylon cord, which is the shutoff to the valve on the bottom of the tank.

I've watched steers bump the float with their nose. It is like pushing down on a bobber. Keep a hot wire close to the tank, this keeps them from chewing on the hose that feeds the tank. The hot wire also prevents them from trying to bully the tank around. I just walk up and dump the small tank, unhook the water hose and move to the next paddock. This is the ultimate water system as far as I am concerned.

I laid 1200' of 3/4" 160 psi above-ground polyethylene pipe from a hydrant that I had set on another farm to a neighboring farm that I had just leased. I did not want to run buried pipe. That way I could roll the pipe up and take it with me when the lease ended. The regular above-ground pipe was 100 psi, which cost 13 cents per foot. For 15 cents per foot, I could buy the 160 psi pipe, which had a lot more wall thickness. It was a tougher pipe for a minimal additional investment.

An NRCS technician ran the pipe size and gallons per minute rate that I had at the hydrant through her computer program. She called me back and said, "Greg I have some bad news for you concerning your 1200' of 3/4" water pipe. According to my computer program and pipe friction table, you will not have one drop of water coming out of the end of your line."

Needless to say, I was sick. I already had all the pipe laid under the hot wire. I just needed some fittings to hook to the hydrant.

I hooked it up anyway and walked down to the end of the water line where I had a quick coupler in place. I snapped on my riser to the quick coupler and water came pouring out. I couldn't believe it. The computer was wrong. I had tremendous water pressure.

The line went 800 feet across level terrain, then dropped down into a bottom the remaining 400 feet. I had 10 gallons per minute at every paddock watering site. I had 12 gpm at the hydrant where I hooked on. The NRCS technician had to come out and see for herself before she believed me. I know the drop in terrain helped. There were also no hills to climb.

That 1200' feeds seven paddocks on a separate farm. Each 400' roll cost $60, which supplies the entire farm with water for $180. That is cheaper than I can dig a pond and gives me tons more flexibility than a pond.

I am very fortunate to live in a water district that has super low water rates (for now). As I said earlier, piped rural water is always cold. Stockers love it.

A 600 lb stocker in my area will drink around five to ten gallons of water per day. When the temperature is below 70 F, a 600 lb calf will drink five gallons or less per day. When the temperature reaches 80-90 F, a 600 lb calf will drink eight to ten gallons per day. Take 30 days x 10 gallons per day = 300 gallons/calf per month. With 100 head of stockers that comes to 30,000 gallons consumption per month.

Most of my farms are split between ponds and rural water. This holds down on the rural water consumption considerably. Even at 300 gallons a month per calf, my water bill for that calf would be 60 cents. That is very cheap for the amount of gain that I am getting from providing them with clean cold water.

There are some water companies that would be very cost prohibitive to hook onto. Check out those water rates before setting a meter or hooking onto an existing meter.

Before there was rural water, most farms had drilled wells. A lot of these wells are no longer being used, because the landowner hooked onto rural water. Check out these drilled wells. They may very well be a cheap endless water source for the entire farm. Some of the older farmsteads have cisterns that are hooked to a house or barn gutters. These old natural stone cisterns can be as deep as 75 feet. A small gas pump or electric sump pump can deliver a lot of water out of a cistern.

I talked with an old timer who used to travel the country digging cisterns. He explained that they would start by laying a round ring of flat stones on top of the ground right where they wanted the cistern perimeter. It looked like a campfire ring. Of course all the rocks were flat though. Once the ring was completed, they started digging the dirt out from under the rocks and also the center of the ring. One fellow would wheelbarrow the dirt away, another would keep placing flat rocks on top of each other at ground level. The whole rock formation would keep settling down as the dirt was removed from under it. They repeated this process until they got to their desired depth. I always wondered how the old timers got those flat rocks to lay together so tightly and not cave in toward the center of the cistern. It also helps explain how they dug a 75-foot hole without the dirt sides caving in on top of them.

Natural springs can be a good clean source of livestock water, if you are lucky enough to have one on the leased property. Again fence off the spring to keep the cattle out of it. Make a small dam that backs up a pool. Run a water pipe from the pool over to a stock tank for cattle. You will need a large reservoir when running large herds of stockers.

Most springs in my area are relatively small. I have one farm that has an earth moving tire that serves as the reservoir for a spring. The side wall is cut out of it on the top exposed side, which gives the cattle plenty of room to drink. One four-inch pipe from the spring comes in the bottom of the tire, which feeds the tank. Another four-inch pipe is set in the concrete that is sticking up flush with the top of the tire. This is the overflow drain pipe. This overflow pipe transfers the constant flow of spring water to the creek, which keeps the cattle from tromping in mud.

The whole bottom of the earth moving tire is sealed with concrete to prevent water from running out. This tank holds 600 gallons of water. It is also freeze proof because it never stops running out the overflow drain pipe. The stockers really like the cold spring water. The earth moving tires are free. All you have to do is go pick them up. Make sure when you pick them up at the tire shop that they are not steel belted. This makes it very tough to cut the side-wall out, with the additional steel wire in them. Use a front-end loader or a hoist to unload them. They are very heavy. The tire tanks are very durable though. They will never rust or rot.

Chapter 10
Fencing Techniques

I can remember the first time I had made up my mind to give electric hi-tensile fence a try. I had just bought my first farm and was struggling to barely make the payments on it. I knew nothing about electric fencing except that when I was small, my Dad would send me out with the weed whacker to knock the weeds and grass off it. I had to do this all summer to keep the horses and cows in. I sure wasn't looking forward to whacking brush and weeds all summer.

A neighbor kept telling me that New Zealand fence chargers were different. You don't have to worry about grass or brush growing up on them. He was right. The next thing I thought was, well if they are that good, I bet they use a lot of electricity. I was wrong again. They don't.

What really opened my eyes was when I accidentally touched my neighbor's fence one day. It hit me so hard that I felt it in my arm sockets for several hours. I could not believe the pain. That was not what I remembered electric fence feeling like.

I personally prefer fiberglass posts in all electric fencing. They cost more, but they will not rust or rot and you don't need any insulators like metal or wood posts. They will not short out like wood or steel posts either. I look at posts as a long-term investment in my business. It's your choice. Buy the best you can afford and you'll never have to worry with them again. Buy the cheapest, and you'll have to check for shorts in your fence constantly.

> *Every dollar you save is like putting gain on a calf.*

I bought the best that I could afford. I found a scrap fiberglass reject rod source that sells them in large quantities. This makes it very affordable to fence with fiberglass. The rods vary from .5 inch to 3.5 inches in diameter. The big ones make great corner posts. There are a lot of fiberglass sucker rods that are pretty reasonable, if you buy them in large quantities. Look around for scrap fiberglass. You may be surprised at what you find.

If you buy fiberglass posts, I highly recommend that you paint the posts before using them. This protects them from UV degradation. What happens with fiberglass posts exposed to the sun is the resin is eroded away by the UV exposure from the sun. This leaves you with solid fiberglass on the surface of your posts. Every time you touch one, it leaves very itchy fiberglass all over you. Also when it rains, the solid fiberglass post surface will soak up water, which causes voltage leakage to the ground.

Lightly sand the area of the post that will be exposed to the sunlight, wipe it down with acetone and immediately paint them with white enamel paint. Lay out about 50 posts on saw horses where they are snug against each other. Just paint the area that will be exposed to the sun's UV rays. The area in the ground is protected. You can hand brush several hundred in an evening. Your posts are a significant investment, so protect your investment by painting them.

I made up a drilling jig by simply cutting a V lengthwise in an eight foot 2x4, right down the center.

This makes a great rest to hold a round rod while you're drilling holes in it for holding your wires. On the 2x4, positioned over the V is a flat steel plate with a single 1/4 inch hole drilled in the center. This steel plate is what holds my drill bit centered while I'm drilling the post. I have five of these steel plates positioned up the 2x4 for the desired levels for five-hole posts.

When you have the correct tools you can put up an enormous amount of fence in a day.

When drilling line-posts, use just the bottom hole. I use a $30 air drill with a concrete bit to drill the fiberglass rods. I am amazed at how well the concrete bit works. I can drill thousands of holes with one bit, no sharpening. I do all my painting and drilling in the winter. When the weather fairs up and I get a lease, I'm ready.

Fasten a piece of polyethylene pipe on the inside of your post driver. This will protect the painted rod surface from being chipped during installation.

They will not need repainting for eight to ten years. I've got some that were painted 12 years ago and still look good. This is a small price to pay to protect your investment in a good quality fiberglass post. One gallon of enamel paint will cover 700 four-foot line posts. I use white paint, because the cattle can see the posts better.

Another advantage you gain from painting them, they look really good on the leased ground. I have received numerous comments on how nice my paddock divisions look. This is awfully good free advertising for new prospective land leases.

I have three common lengths that I cut and paint. The

four-foot post with a single hole is used for line-posts. The four-foot post with two holes drilled twelve inches apart is used as a perimeter post on all property that does not touch a road. The five-foot post with a single hole is used to hold the wire down in low spots or at slight bends in the fence. Six foot posts are used as three, four or five wire perimeter fence post.

For paddock division single hi-tensile wire divisions, I use a 76-inch fiberglass post, 2 to 3.5 inches in diameter. The post is driven in the ground 42 inches, which leaves 34 inches sticking out. The hole to hold the wire is two inches from the top. This gives me a wire height of 32 inches, which I think is a perfect nose height for most stockers. I can pull a mile of wire from this one corner post. It just plain doesn't budge or flex.

My Dad made the mistake once of getting his truck fender hooked on one as he passed the gate. The whole back bumper and rear fender looked like you took a giant can opener and sledgehammer to it. I figured from the looks of his truck, that I had better go check out my poor corner post. There the post proudly set, still square, as tight as if it had grown out of the ground. The only damage that I could see was some chipped paint on the side of the post.

Some mischievous kids decided to plow through my perimeter hi-tensile three-wire fence one night. They were probably drinking, wanting to have a little fun with the 4-wheel drive. This fence was along a gravel road that had no houses on it beside one of my leased farms. They went through all three wires, but got their front drive shaft hung on all three of the high tensile electrified wires.

A good fence charger will kill any green material touching the wire.

The next morning, I found their front drive shaft still hooked to my wires, just popping like the dickens. They must have had 200 feet of wire wound around the drive shaft. The fence had 7200 volts running through it when they decided to plow through it. The corner post was unfazed. The electric cutoff switch was 100 yards up the road from where they were hung, but they didn't know that.

There were two sets of tire tracks where they had spun in the middle of the gravel road trying to break either the wire, or corner post off to free themselves. You could see where they kept backing up and getting a run at it, to try and break free. I would have loved to have been a little mouse sitting on the corner post, watching and listening to them. Can you imagine the terror of being hooked to 7200 volts and not being able to get loose, just wanting some relief? I bet they sobered up pretty quick, when everything they touched shocked the heck out of them. You probably could have fried an egg on their truck engine by the time they got loose. Sometimes plain fair justice can be so sweet! I can't help but chuckle a little bit, every time I think of the incident.

Install several cutoff switches in your fencing system to help in locating shorts.

Wooden posts are very adequate for corners also. Determine where all corner posts are needed by walking the farm and studying an aerial map of the farm. Use a tape measure to place the location of all gaps and watering sites. I mark each post site with a small red flag.

Hire a tractor with post hole digger to dig the holes for you. Do not waste time trying to set the posts as you dig them. When I use wood posts, I use posts that you can get free for the cutting. Lots of landowners will let you cut posts for a share of the posts, or just to remove the trees from their pastures. I rate wood posts in this order for longevity:

1. Hedge
2. Black Locust (This does not include honey locust, they rot.)
3. Mulberry
4. Cedar

These are the only four in my area that are available. They all work fine for corner or line posts. Insulator manufactures have made some huge inroads on electric fence insulators in recent years. I prefer the black plastic corner insulators with the metal reinforcement strip molded into them. If livestock or deer run through the fence, the insulator will normally take the impact without breaking. The PEL insulators that I use run from 50 to 75 cents each, and are very durable.

If I'm pulling more then two hi-tensile wires, then I use a floating brace as a corner. They work great and take very little time to put in. Before using trees as corners, make sure it is okay with the landowner. Never nail to a marketable tree. When you use a tree, protect it from the wire with a poly tube, piece of tire, board or garden hose. Allow plenty of room for the tree to grow when you are fastening the wire insulator to it. I've had trees swallow up the wire and pull the insulator right up against the bark. Then you have a short.

I've seen people use ree-rod with plastic insulators fastened to them as line posts. These are probably the cheapest posts you will find. Sometimes the cheapest is not always the best in the long run. Remember, every steel rod or fence post that you incorporate in your fence could become a potential ground rod. If something runs through the fence and knocks the insulator off the post, then the wire is lying against the steel post. Nothing kills an electric fence quicker, other than unplugging it.

Old wire fences are a headache.

Old wire fences will also give you headaches, particularly if you have placed your high tensile electric wire fence right next to one. What happens is the deer will jump your new fence and land in the old fence, which may throw a piece of old barbed wire onto the electric fence. This is where I find most of my shorts in the fence. Most of the time the other end of the barbed wire is buried in the dirt, which makes a heck of a ground rod.

Ask yoursef: Do I really need this? Is there an alternative that will work and is free?

I have farms that are fenced in all fiberglass posts. These are the farms that I literally have no shorts with. I may find several wires knocked off the posts, but this will not short out a good charging system. I have small trees, multiflora rose, briars, etc. growing up against my fence in places. I don't worry about it. None of these short out the fence.

A good fence charger will actually kill any portion of green material touching the wire. This is why I keep all power on the fence at all times to kill any new green material that grows up and touches it. If I shut off all sections of the farm that are not being grazed, then these sections of fence get a mass of new green growth covering them. When you come back through in 30 days with your grazing rotation and energize this wire section, then you have got a tremendous load on your fencer. The fencer has to start from scratch, killing all the new green accumulated growth that is covering the wire.

I will install several cutoff switches in a grazing system, just to help find shorts. If you have a short in the fence, start at the back of the farm with your voltmeter and start turning off switches. When you get a good voltage reading, you have found what section of fence the short is in. This can literally save you half of a day walking fences looking for the short.

Another advantage of several switches is when you're working on the fence you don't have to go clear to the charger to shut off the power. Don't get carried away with them though. They are expensive. Four to five switches in most instances would be adequate for a 200 acre grazing system. Fencing dealers will tell you that you need a switch for every paddock, but you don't. Remember, you're trying to keep the money in your pocket, not line everybody else's.

I have come up with a homemade switch that works fine for paddock divisions and other short fence divisions. I take a 5' long, 1/2" diameter fiberglass rod and drive it in at the gate end that does not have the power feeding it. Then I hook a high tensile wire with a piece of poly tape together for visibility to the gate handle. The 5' fiberglass rod acts as the "spring" in your gate. It also helps make a good tight connection to carry the voltage to the paddock.

If a calf runs through it, there is no wire spring to stretch into a worthless 40' gate. The 5' posts will bend with the impact and return to the upright position. Make sure the gate wire is fed from the cold side of the fence. This way you're not handling a hot gate wire. I love these gate switches. They work great and are cheap to build. I have never found a broken one yet. The calves don't mind that they are homemade.

For gates, I use half-inch-wide white poly tape with a piece of insulated underground wire as a handle. It is a lot cheaper than a $15 spring gate, which usually ends up 40 feet long after a cow runs through it. Smile and keep your money in your pocket. The poly tape gates work great because the cattle can see them. I've never seen a trained steer run through one. They are scared of them. A 660 foot roll of white poly tape will cost you $22.00. That will build 41 gates 16' long, which comes to 53 cents per gate.

> *Remember, you're trying to keep the money in your pocket, not somebody else's.*

The poly tape will stretch some when you close the gate. This keeps it extremely tight.

You will have numerous gates in your grazing system, so if every one of them costs you 53 cents each, rather than $15.00 each, that is a major savings.

Every dollar you can save is just like putting gain on a calf. It's money that is in your pocket, instead of someone else's.

You're probably thinking this guy is a cheapskate. You need to be this way. It gives you a big edge when finances are tight. There is no quicker way to ruin a young grazing operation than to go out and build a huge overhead that the grazing operation can not service. The next time you walk into that farm store to buy supplies, ask yourself, "Do I really need this? Is there an alternative that will work, that is free?"

Let's get back to fencing, I'm sorry I got carried away with spending habits.

Around all watering areas I use white poly tape around the tank float area, hydrant, or coupler. This keeps the steers from messing with it. It is a very visible predator to them. They want no part of it. You can lay it on the ground without voltage on it to serve as a quick fence when moving them. They will not step over it. Watering areas get a lot of pressure. A strand of poly tape will keep them from busting down an adjoining gap that serves another paddock.

I use one 12.5 gauge hi-tensile wire in all of the paddock divisions. It is strong enough that if a deer runs through it he may knock the wire off the post, but it will not break.

Electric fencing is the most powerful tool in a MiG system.

You must have access to a spinning jenny wire dispenser for your hi-tensile wire. The jenny holds the roll of wire under control and has an adjustable tension to let you regulate how much force it takes to spin the wire off.

It is amazing the fence you can put up in one day with the correct tools. You will need access to a set of wire crimpers. Use the wire crimps that completely seal down around the wire when you crimp. I started out using the steel flat crimps to fasten all joints. That was a big mistake. The flat crimps will not hold up when something heavy runs through the fence. The wire will always pull out of the flat crimp.

Use quality materials for your electric fence and you'll sleep well at night.

Use in-line strainers to keep all wire tight. Watch for annual sales on in-line strainers. I have found them for $1.00 each. I don't use springs. They are expensive and you really can get by fine without them.

With all other non-road frontage perimeter fence, it depends on the type of livestock you are planning to graze. With cattle I run two wires on all perimeter fence not touching the road. You can space four foot posts 40 to 60 feet apart on all interior fences with one-strand wire held to the post with a five-inch cotter key. With sheep and goats, run three to four wires on all perimeters not touching the road. For interior fences run two strands.

In all fencing, when I have low spots that need the wire held down, I will drive a 5' long, 1/2" diameter fiberglass rod at an angle, four feet in the ground. This rod will have one hole drilled in it at the top of the rod. Run a wire through the hole and pull the fence down to the rod. By driving the rod in the ground at an angle, the wire is putting a cantilever pull on the rod, which prevents it from being pulled out of the ground.

Another alternative for holding wire down in low spots is to use a helical earth anchor with an insulator tied to it. Then pull the fence down to the insulator that is secured to the anchor. I have never had one pull up. It's very stable and simple.

When pricing wire and posts, check around for the best price. It will save you some serious money. Always buy a minimum of 170,000 psi 12-1/2 gauge hi-tensile wire. Beware of some hi-tensile wire that is extremely cheap. My wife bought two 4000' rolls of hi-tensile wire on sale at a local farm supply outlet for $40.00 per roll. She thought she had found the ultra bargain on wire. We started stringing it up and found out that it was very, very brittle. It would break every time you made a connection. It was a real undertaking to wind it on an in-line strainer. It actually chipped my side cutters when I cut it. We finished that roll and brought the other new roll back to the store.

Beware of hi-tensile wire that is extremely cheap.

Buy a good high voltage, low impedance charger and put in a minimum of three six-foot galvanized ground rods. I cannot stress this enough. You should have a charger that will constantly be capable of maintaining 3000-6000 volts. I get kind of nervous when a charger drops below 2000 volts. I like to hear a pop when a steer touches a wire. It should sound like a .22 rifle going off. Then you can sleep good.

I helped build 12,000' of perimeter fence between a newly leased farm and the landowners' neighbor. The neighbor volunteered to energize the new fence between us. I asked him what kind of fencer he had.

He replied, "I bought the biggest baddest fencer they had at the store. It is putting out 10,000 volts on the meter peg that is mounted on the front of the charger."

I was impressed, but I'm from Missouri. You have to "Show Me." So I asked him if I could see it working.

He brought me out to the barn and I was horrified to see his setup. His ground wire ran out of the barn and was hooked to a two-foot piece of rusty ree rod. He had a plain hi-tensile non-insulated wire going through an old rickety pig barn, that was hooked to this worthless brand charger. This feed wire was within an inch of nails, cattle panels, wire, etc. The charger was dangling from one old rusty nail. Directly below the charger was a pig mud wallow. You could hardly hear the charger clicking, but every time the charger clicked, the little needle on the front would travel over to 10,000 volts.

I hooked up my digital fence tester to the wire coming out of the charger and couldn't even get a reading. I told the landowner, "Hope you didn't throw the box away that your charger came in. You might want to think about bringing it back."

He didn't believe my tester was reading correctly, so he reached out and grabbed a hold of the wire. He couldn't believe it, and replied, "They told me that I bought the biggest and strongest charger they had in the store."

I told him to go see my fence dealer. He would sell him a real charger for about the same money. I also instructed him to pick up three six-foot galvanized ground rods. His new charger system is kicking out 5600 volts on 12,000 feet of wire. He is a happy camper and I don't have the worry of our stock getting together anymore.

For leased land, hi-tensile electric fence is the cheapest and easiest to install.

People are sometimes skeptical about using electric fence, because they assume they use a lot of electricity. I have four farms that have nothing else on the meter except an electric fence charger. After a full summer of grazing, the meters on the four farms have yet to turn over one kilowatt-hour. I know it is hard to believe that something that is so hot doesn't use any kilowatts. It still amazes me.

I'm absolutely convinced the cheapest and easiest way to go on leased ground is to install hi-tensile electrified fence. It's easy to put up, easy to take down, and has very little maintenance if correctly installed.

I don't use electrified barbed wire of any kind. It is mean to put up and even meaner to take down. You will have more holes punched in you than your animals will. Barbed wire is meant to be strung in place and left for it's remaining life, period.

I installed a two-wire perimeter fence on a back 40 leased farm in one day. This included chain sawing out the brush where the fence was being erected. This also included driving several fiberglass corner posts. With the spinning jenny, you can string smooth hi-tensile wire as fast as the ATV and ground terrain will allow you to travel. If I had decided to fence that back 40 with barbed wire, it would have taken me weeks to erect the fence. It would have also cost me three to four times the dollar amount it took with hi-tensile wire.

I knew a fellow who had come up with all kinds of new money saving homemade gadgets in running electric fence on his farm. They ranged from using steel reerod for posts to rubber inner tubes for insulators. With electric fence materials you get what you pay for.

Electric fence has very little maintenance if correctly installed.

Several years later, the same fellow swore that electric fencing was not a reliable fence at all. His cattle had gotten through his electric fence and ruined the neighbor's cornfield. He literally could not keep them at home. He tore out all of his electric fence and swore that he would never string another inch of the nasty stuff. If only the poor fellow had used good quality fencing materials at the start, he would have never suffered the losses of paying for his neighbor's damaged crops.

I cannot even visualize trying to set up a MiG farm with out electric fencing. It is the most powerful tool introduced to grass farming in the last decade, as far as I'm concerned. I think I've made my point. Use quality materials when setting up electric fencing and sleep well at night.

Chapter 11
Improving Your Existing Forages

Most of the farms that I lease are idle, open, fenceless ground. They may not have had anything done with them for years. Usually a huge accumulation of old dead thatch is built up on the soil surface. Most idle pasture is also void of legumes, so the first thing to concentrate on is getting them in the grass sward.

I have found for the most economic stocker gains, you need 20-40% legumes in your grass stand. The legumes have the unique ability to fixate nitrogen from the air and store it in their roots for later use. The grass roots come along and steal the nitrogen from the legume roots. You don't need to add any commercial nitrogen if you can keep a good stand of legumes in your grasses.

There are several reasons why there are no legumes in the idle pasture. One is because the dead thatch has covered the ground so effectively that the sun can not touch the soil surface. Another reason is a heavy grass sod is too competitive for a young clover stand to compete against. You have to hurt the existing grass stand, either by heavy grazing, light disking, burning or a combination of these. Take a soil test to see what kind of pH you are dealing with. If the soil pH is above six you should have no problems establishing legumes.

You will never get legumes established if they have to compete with a mature grass stand and heavy dead thatch. I purposely try to overgraze the grass stand going into winter. This will give your legumes a better chance of establishment in the spring.

If you decide to burn the old thatch, you should attend a burn workshop first. When I attended my first burn workshop, it really opened my eyes as to the correct way of conducting a controlled burn. There is no sicker feeling in the world than to watch a fire take off in the opposite direction that you intended to burn. Once you light that match, you are 100% responsible for it's destruction. I would also strongly recommend some form of pressurized water. When you have a wind direction change or it jumps a fire break there is no replacement for pressurized water. I use a 25 gallon sprayer tank on my ATV, a very mobile unit.

I was burning a newly acquired leased farm one afternoon by myself. I had my 25 gallon pressurized water tank sprayer on the ATV and was watching a lazy grass fire burn down the ridge. There was a slight wind just barely pushing the fire. When the fire got to the bottom of the ridge, it started up the other side. The wind had quit. I felt very confident that I had control of the grass fire, when all of a sudden it started whooffing and blasting up the opposite hill sideways. It was generating its own wind from the heat of the flames by going uphill. I immediately attacked it with my spray rig, hitting it right at the base of the flame. I had about a quart of water left in my sprayer when the last flame was extinguished. That one burn made a believer out of me on pressurized water. A fire will burn ten times faster going up a hill than down a hill.

A newly burned pasture makes a beautiful seed bed to

broadcast clover into. The ashes make the ground just suck up the legume seeds. The blackened ground warms up faster and lets the sun go to work on those new clover seeds. It is also very easy to see on the burned ground where you have spread with your seeder by the tracks.

I use what grass is present, and control the grass sward to give the legumes a chance to express themselves. Do not go out and spend money on new species of grass just because the seed salesman says that this particular grass is the best thing since sliced bread. I did a complete pasture reestablishment on a piece of ground several years back when I was stupid. I put down brome, timothy, orchard grass, red clover, alfalfa, birdsfoot trefoil, and alsike clover. I had read that the more you put down the better. A successful grass farmer told me right after I got finished that in four to five years I would have nothing but a stand of fescue and red clover.

A freshly burned pasture makes a beautiful seed bed for broadcasting clover.

I informed him that I did not plant any fescue and that the birdsfoot trefoil was going to be my dominant legume. Well guess what? He was right when he forecasted my grass stand. I had spent a ton of money on all those different grasses and legumes. I ended up with fescue and red clover. The successful grass farmer had also told me at the time that I had better learn to manage fescue and keep a vigorous stand of clover in it. This was the only way I stood a chance of making any money grazing cattle.

At the time I thought he was kind of a pessimistic fellow, but I learned he knew exactly what he was talking about. The ground where I did a complete pasture establishment took several years to establish the sod back, which I had tilled under. I had some severe pugging even with the light stockers. I had very little sod, so running anything in the fall was out of the question. When you are young, you have the attitude that you know everything already. I know I did. I wish on hindsight that I had listened to the older successful graziers more. It would have saved me lots of money and years of bumps.

I never do a complete pasture reestablishment on any ground anymore. It takes a lot of money and too much time to build up a sod that you will not pug when you graze it. I do not want the ground out of production while I'm paying to lease it.

Keep in mind when establishing legumes in permanent grass stands that the young legume seedling roots have a battle going on under the ground as well. If the present grass stand is an aggressive stand of fescue, well then the little legume seedlings have to compete with the mass of fescue roots as well as the grass canopy above the ground. This is why it is so important to physically hurt the existing grass sod. If you graze it hard into late fall, you have hurt the grass roots' energy reserves. The grass stand will be a little slower getting started in the spring green up. This will give the legume seedlings a window of opportunity to get established.

I try to have an agreement on the contract with the landowner to deduct the lime, P&K off the lease if I agree to pay for it. This really jump starts a piece of marginal pasture. I have found a lime hauler that will spread one ton of lime to the acre, at no extra cost. Look around and compare prices for getting lime spread. It is better to put a little lime down each year or every other year, than it is to put down three to five tons per acre every six to eight years. By putting down some lime every year, you keep the top one to three inches of soil from getting too acidic. This really favors the legumes as well

Lime really jump starts marginal pasture.

as better growth in your grasses.

Sometimes if there is a very heavy stand of fescue after it has been grazed off, lightly disk one pass over the field to ensure good seed-to-soil contact. It may look like you are not tearing up much dirt, but you are disturbing the soil enough if you use an adequate wheel disk. You can also no-till legumes into the existing grass stand with good success. The pasture should look like a golf course before you seed or you have too much thatch left.

The pasture should look like a golf course before you seed.

I inoculate all clover seed the night before in my garage. I lay out a plastic tarp and spread the clover seed out to where it is 3-4" deep. Then I sprinkle the top of the clover seed with sugar water, (1/2 cup sugar to a gallon of warm water). I will knead the clover seed until the entire pile is just moist. You want the seed just moist, not dripping wet. Then sprinkle the powdered inoculant over the top and work it into the moist seed. Turn the seed once, right before you go to bed. The next morning you have a pile of seed with cured inoculant covering the seed. I get tremendous germination with this method.

A red clover plant scientist told me about this method. His theory was simple. If you're going to go through the expense of buying the clover, broadcasting, rolling, discing, culti-packing, etc, then why not give the clover its best chance at having a productive life. My seedlings just seem to explode out of the ground when there is adequate moisture. Frost seed three lbs/red clover, per acre, in late February over the existing grass stand.

You can also put down one lb of ladino clover, but be careful you don't exceed that or you may have bloat problems.

There are 800,000 seeds per pound of white clover.

Mark off an acre, weigh out a pound of seed and set the seeder (I use a Herd broadcast seeder) to 1/16th of an inch. I step off a 208'x 208' square (acre) in the pasture with markers. Then I broadcast the acre with the new setting. Look in the seeder after you have covered the acre to see how much seed is left. Adjust your seeder from that point. This method works well with any type of seed.

I cannot stress enough the importance of getting legumes established. When it gets hot and dry in July and August the grasses go dormant and the clovers stay green, which makes some very nice forage for stockers to graze. Lots of other nice things begin to happen with the addition of legumes. I have noticed the deer and turkey are really drawn in by the presence of legumes. The turkeys feast on the bugs that are attracted to the legumes. The deer nip the tender tips of clover, which builds nice racks. Quail come in for the insects and seeds as well. Rabbits love the tender clover leaves also.

I've no-tilled three to four dollar a pound alfalfas on some of the leased farms. This was a mistake. The fertility is not high enough to get a profitable stand. I wasted a lot of time and money for no return. I would have been better off concentrating on red clover, period. The alfalfa is more bloat prone anyway, but I was after those magical two to three lbs of gain a day that I had read about. Most of the farms that I lease are low on fertility to begin with, so I strictly concentrate on what will grow in that environment with some lime and phosphorous added.

> *Inoculate all clover seed the night before sowing.*

Be patient when building your forage base. Sometimes it takes three to four years to really get things showing a noticeable improvement. Patience may save you some money also. When you make a management change, give it some time to see the results. You can tell pretty quickly whether you are going forward or backward in your pasture development.

I agreed to winter 85 custom grazed heifers on a leased farm one fall. I bought all the hay at $20 per 1200 bale including delivery. I concentrated all large round bales on paddocks that had the worst fertility. Those poor fertility areas were solid broomsedge grass. The broomsedge was so thick that it looked like wheat fields ready to combine. The bales were placed on high spots in the broomsedge fields and fenced off with poly wire.

Be patient when building your forage base.

When it came time to feed them, I used my Toyota truck to unroll them across the fertility-starved paddocks. I simply drove a sharpened 1" steel bar through the center of the bale, which is where the hay is not as tight. I hooked a log chain to each end sticking out of the bale. My truck had no problem unrolling them like a bobbin of thread.

While unrolling it, there was a cloud of seeds being spread across the pasture also. It didn't matter if I hooked onto the bale in the opposite direction that it was rolled up, I just got a longer wind row of hay rolled out. The manure distribution was awesome. It was a solid carpet of manure where the bale had been. No pugging was done like you have with feeding out of a big bale ring. By unrolling the bale all 85 heifers could get at the bale. Very little hay was wasted. By the next day all you could see was the row of manure where the bale had been unrolled. The best part is the heifers spread all the manure and urine out on the pasture where it could do some good.

The following spring the pasture just exploded with dark green lush grass in the bale fed areas. The earthworms were having a hay day on the old remnants of manure and hay residual. There was a multitude of different species of grasses

Chapter 12
Utilizing Every Green Growing Leaf

What your goal should be as a grass farmer is to have as many solar collectors (grass leaves) available to the sun, as many days of the year as possible. You are actually collecting energy from the sun and converting it into meat. If you graze the grass off too short, you have severely reduced the amount of energy that the grass leaf can capture, therefore reducing the amount of meat that could have been produced. If you let the grass get too tall, then it is mature and the livestock will not eat it very well. Weight gains will suffer.

I have found that in the Midwest, there is a two to three week window in the spring where the grass grows so fast that you can hardly keep up with it. This is why it is so important to graze it hard in April so that you have a chance of keeping the seed head from forming and the grass stays in a young vegetative state.

Move the stock twice a day. Don't worry if there is

grass left ungrazed. You are trying to get over the entire farm as fast as you can. As the forage slows down growing, move the stock every two days. Remember to be flexible in your grazing rotation.

Watch your manure piles. If they are big thick pies, the stock are consuming some very unpalatable grass. The manure piles should resemble a flat saucer (very palatable), which tells you that the stomach of your stock is digesting the forage very quickly and the animal is gaining weight.

Be flexible in your grazing rotation.

Turn out stock on grass when you can see green. If you wait until it is four inches tall, you will never catch up to it.

If you let the grass get mature, you have wasted it.

Remember the grass plant is trying to put on a seed head. You as a grazier are trying to keep it off. If the grass gets away from you, clip it off leaving six to eight inches by the end of May. It will grow back vegetative and the stock will eat it. Usually by the end of May there is still enough ground moisture that the grass will grow back fairly quickly. Clip paddocks right after you take the stock out. That way it has a full rotation to grow back before it is grazed again.

If possible, stock pasture heavier in the spring. Cull out the heavier stock after the grass slows up going into summer. This will give you a better mowing machine when the grass is growing the fastest. The more flexible you are, the more successful you will be. If your grass gets ahead of you, drop several paddocks from your rotation and stockpile these to graze later in the summer when it gets dry.

I had a farm perfectly stocked one spring. I had some heavy yearlings that just mowed the paddocks every rotation during the spring. When you moved them out of the paddock, there was a 2-3" residual left. These steers grazed from April 1st to July 21st and absolutely kept almost every seed head off the grass. The pasture had about 50% red clover in all the paddocks. The steers looked like they had been on grain. They were big blocky animals. The pastures were allowed to rest until the middle of October, then I brought in a new bunch of calves. No nitrogen was needed for fall grass stockpiling because of the strong stand of red clover. It made a very nice fall pasture to turn calves onto. This is the ideal scenario that you should be shooting for on all your farms.

I'm convinced that most cattlemen feed entirely too much hay in their livestock operations. I look at all my pastures in July and August for building up a reserve of standing hay that the calves can harvest. The calves are a lot more efficient harvesters than any hay making machinery that I know of. They do not depreciate like all that heavy metal hay machinery either. The calves are putting on weight, making you money. It is really amazing how hard we have worked at taking the animal out of his natural environment of grazing. Think about it. Cows grazed and propagated for thousands of years without all of this fancy machinery that we think we have to have to keep ole bossy alive through the winter.

Every time you fire up that tractor to make hay there is a possibility that something is going to break, puncture or blow up. The old saying goes, "If you're going to listen to the music, you got to pay the fiddler." Well this perfectly describes what is going to happen when you start using heavy metal constantly around your farm. Next time you hop on that tractor seat to go make some hay, ask yourself, "What would I have to do to graze this same field in December and January rather than baling it into worthless hay?" Taking time to answer these types of questions are what will make your farms profitable.

Turn out stock on grass when you can see green.

You need to concentrate on eliminating as much hay

feeding as possible. Keep a 30-45 day emergency supply for a sudden extended storm if you plan on wintering any stock. I will graze all through the winter as long as the ice does not freeze thick enough over the forage, preventing the calves from getting it out. I graze the areas of each farm that are the hardest to get to early in the winter. Save the more accessible areas of the farm for later in the winter when the weather is worse.

Most cattlemen feed entirely too much hay.

Missouri had the worst winter on record in 2000. Starting the first week of December, we got 8" of snow that was followed by two to three days of freezing rain. The ground was completely sealed off to any critter that needed a meal. Then the temperature dropped to below zero and stayed there for three weeks. I've never seen a winter in Missouri that stayed just bitterly cold for so long without any reprieve. I remember watching the weather channel and it was warmer in Canada then it was in Missouri.

During this period I was custom wintering 170 stockers that I had fall grazed. I had agreed to winter them so that I would have a group of trained big calves to dump on my spring rush of grass. I always seemed to lose my grass in the spring, because the little 400 pound calves didn't lop off big enough bites. I was always getting the newcomers trained to move, while the grass was getting ahead of me by the minute.

I had 85 stockers at two separate farms, with 50 purchased large round bales of hay strategically placed for each group. Both farms had a nice fall stockpile of grass on them. I had calculated that I had enough standing stockpile to graze until February without feeding any hay, assuming there was no heavy ice. Well, I got the ice, snow, and bitter cold weather all in one nice neat package.

Through the month of December I fed 27 bales of the

50 that I had allotted for each group. By January 5th, all snow and ice was melted off. I went 75 days without feeding anymore hay to either group. I rotated the stockers through the paddocks of stockpiled grass. They loved grazing, if it was available, rather than eating dried hay. The stockpiled fescue was still partly green and the calves went after it like candy. I fed some pelleted corn gluten to each group every other day for their protein requirements. The calves built a nice frame over the winter, not fleshy at all. They flat poured the weight on when the spring rush of grass exploded. This is the kind of gain you want to maximize – cheap grass gain.

With purchased grain gain, the feed store will give you one of their pretty jackets, but that doesn't do much for the big gaping hole in your wallet from buying all that feed. You have to minimize your cost of gain, or you might as well stay in by the warm stove all winter with your feet propped up. You can not starve a profit out of calves, but you can choose when to put weight on and when not to.

Winter is not the time to try and put lots of gain on calves. It will literally break you. All you're doing is giving all the profit to the feed company. I don't have a problem with them making a living, but not at the expense of me going broke.

Concentrate on using your pastures to graze in the winter, any time there is not snow cover. Just maintain the calves' weight through the winter, then pour the green grass to them in the spring. It's pretty simple stuff really.

Remember, there is always money to be made grazing green growing leaves of grass, so you want to be able to grow as many of them as possible and harvest them with a growing animal. When we get away from this thought process is when the bills start growing and our profits disappear.

Chapter 13

Controlling Costs and Saving Money

The best way to control your costs is not to spend any money on anything unless it will pay you back double. Concentrate on water, lime, P&K. Graze large groups of livestock. It spreads your cost over more animals, which lowers your operating costs per animal. It also doesn't take much more time to move 200 head through a gate than it does to move 20 head. I strongly suggest that you utilize every resource that is present on the land first. Do not invest a lot of capital in materials until you have exhausted all present resources.

I do all the grazier-related work possible myself, which holds down on labor costs. You need to be innovative. Things do not have to have a factory sticker or paint on them to perform.

Before you buy that next item for the land ask yourself, "Is there an alternative to this item?"

Resist the urge to go out and buy that new 4-wheel drive diesel truck. You don't need a new diesel 4-wheel drive truck to be a successful grass farmer. Buy a used small truck. It will get you from farm to farm as well as that new fancy truck.

With custom grazing you don't need to own a trailer. Cattle owners are responsible for hauling all livestock to and from farms. If you own the livestock, hire the hauling done. A stock trailer is a huge investment for a new grazing business. Then if you have a stock trailer, you can no longer get by with the small pickup. It takes a large pickup to pull any type of stock trailer loaded.

If you decide to purchase an ATV to build fence and move cattle with, then buy a used ATV. They will perform as well as the new ones and you can keep the seven thousand dollars in your pocket. The livestock don't care what you putter around on.

I see a lot of young ranchers buzzing around on the biggest and latest version of 4x4 ATV wheelers. Think how many steers you could buy with that money? How many quality water systems could you put in? That new wheeler is a classic example of a "Boy Toy" that you have to train yourself not to purchase. It robs you of your hard earned profits and doesn't return you a penny. When the next new big ATV version comes out, then you decide you have to have that one. It is a vicious cycle that destroys your wealth. It's not easy to see your buddies cruising around on that new shiny ATV, while you're puttering around on your old one. Learn to smile and say to yourself, "Look at all those steers he didn't get to buy."

Another wealth destroyer is the farm tractor. Grass farming does not require a tractor. I do own an old Allis Chalmers WD-45 tractor that I bought 23 years ago. It is a 1949 model and pretty simple to work on. If it doesn't want to start, it is either points, fuel or a dead battery. It's a pretty simple old tractor. I use the tractor to haul in my firewood out of some pretty rough woods. I don't feel bad if it gets a few scratches on it while hauling out wood. I feel guilty for owning the tractor, because I only use it three to four times a year.

They say if you don't use the tractor every day, you should not own it. Well I'm guilty of that crime for sure. Maybe my conscience will make me sell the old tractor one day.

This $600 WD-45 is a far cry from what most cattle operators have on their farm though. You see numerous 30-50 cow operations that may have several tractors ranging from $6000-$20,000 each. It takes a lot of grass gain to be able to pay for that kind of heavy metal. Some cattle operators will have a mowing tractor, rake tractor, and baling tractor. This way they don't ever have to unhook any implements.

Not only are tractors expensive to own and operate, they are very dangerous to operate. I personally know several neighbors who have been killed or permanently disabled by them. This is one reason I feel there is a tremendous opportunity for our youth to learn grass farming. It does not require heavy metal, which makes it safer than traditional farming. Learn to love the ownership of animals, forget ownership of heavy metal. Your bank account will love you to death for it.

Keep a daily log of what you spend and what you spent it on. This is an example of how simple it is to keep accurate detailed records of your expenditures.

Date	Description	Cost
1/2/02	Wire, nails, posts	$15.00
1/7/02	January electric meter	$8.00
1/12/02	100 lbs livestock salt	$4.00
1/28/02	Annual lease payment on Smith farm	$200.00
1/29/02	Fuel for ATV	$5.00
	January Total	$236.00

At the end of the year you can categorize the items together. This gives you a real good blueprint of where your spending is taking place. Maybe there are some areas you can work on to reduce spending.

There are lots of other ways to keep track of your costs

besides the example above. I like the one above, because it is so simple. It is wonderful at tax time to be able to calculate your farm expenditures for the whole year in two hours.

By having an accurate description of each item, it makes it very easy to put them in their correct category for the tax accountant. My tax lady loves to see me come in. All my expenditures are in their correct categories. She fills in the blanks and I walk out. It used to take weeks to get all the correct figures so that I could send off the taxes. Those days are history.

Your standard of living should reflect your grazing operation, practical but not exuberant.

By keeping accurate monthly records, it really keeps you focused on not overspending. You spend more time thinking of ways to avoid spending money on items that you think you need, but really can get by fine without them.

Next time you decide to make a major purchase for your grazing operation, think about it for several days. Ask yourself these questions:

1. Will it increase my grazing operation's NET profits? If yes, estimate how much?
2. What are the results if I don't make this purchase?
3. Is there another alternative that would be more cost effective?

Once you have convinced yourself 100% that the purchase is going to greatly benefit your grazing operation, check around for the best price on your item. The same item can be priced 20-50% higher at one location vs. another. Take the time, do your homework on comparing prices, and it will save you some serious money.

A perfect example of mulling over a major purchase decision was when I bought my farm. I simply cannot believe how stupid I was. I decided when I bought my farm that I needed a brand new 50-60 horse power John Deere tractor to do all the work on the farm. I had two different John Deere dealers rolling out the red carpet for me. They rolled out a whole line of new tractors to let me test drive.

Thank god I finally came to my senses. The more I thought about the reasons that I needed to own a new tractor, I could not justify a $20,000 tractor. At that time I could have bought seventy-five to one hundred stockers for the same price that I was going to pay for the tractor.

The more you can save, the quicker you can take the bank out of your operation.

A lot of my farming friends had new tractors at the time. I was like a cow following the herd. Guess what? A lot of those same friends are not farming in any form today.

Your standard of living should be a direct reflection of your grazing operation, practical but not exuberant. If you're spending more than you're earning, you're swimming up a creek. You're setting the grazing operation up for failure, because the farm will be a mirror image of the way you live.

Your goal should be to become the lowest cost grazier that is possible for your scale of operation. If you are the lowest cost grass producer, it gives you an unfair advantage against the competition. Their costs are higher, which means their profits are lower.

There is more money to be made in your grazing operation by cutting spending, than implementing any new technology in your operation. It is simply amazing the money that can build up, if you don't allow yourself to spend any. This is exactly how I paid off my farm and house in three years. I never carried any cash in my wallet. I figured if I didn't have any cash,

I couldn't spend it, my wallet was empty.

You're probably thinking by now, this guy is nuts, maybe! I absolutely saved every cent that the leased farms generated and paid off all my debt. It is a wonderful feeling to be out of debt. I highly recommend it.

The closer you get to paying off your debt, you'll start to see some daylight. Then you get even more hungry to reach the light even quicker. It is almost the same as tempting a hungry dog with a fresh raw juicy steak.

Another way of controlling your costs is to stay at home. If you go to town, you're more than likely going to spend some money. If you stay at home, you're making money because there is nothing to spend it on. I know this sounds pretty tight fisted, but try it for a month. Every time you think you need to go to town to pick up something, resist the urge.

Remember, the more you can save, the quicker you can take the bank out of your operation. Your profits should be invested in a growing class of livestock. Your goal is to own your livestock. This keeps the bank from telling you when to sell them.

Chapter 14
Resisting the New Truck Urge

You're probably thinking right now, "What's buying a new truck got to do with grazing?"

My answer is, "Your whole grazing livelihood depends on this one particular decision."

You're thinking, "This guy is nuts, and now he has confirmed it."

Well maybe so, but there is no hope for you to ever develop a profitable grazing company if you have the "New Truck Syndrome." That is putting it pretty bluntly, so let's dig into the topic further and uncover the grizzly details.

I heard a true story about Gordon Hazard, a Mississippi stocker grazier, concerning pickups. Gordon Hazard is one of the most successful stocker graziers in the world. Gordon was driving his old pickup across his pasture to move a herd of stockers.

A young man who was riding with him asked, "Gordon why don't you buy a new pickup and get rid of this old truck?"

Gordon sat there a minute and replied, "I can't afford to

son." This is the mindset that Gordon uses to manage his stocker operation also. His philosophy is, "If it rusts, rots, or depreciates, don't buy it." The stockers don't care what he drives. They're happy to see him because they know they're going to a new grass pasture.

I believe this is the prime reason he is so successful. He grazes 1800 stockers a year by himself and he is 80+ years old! He is 100% self financed also. How many large graziers can say that? He did not accumulate his wealth by buying a new truck every two years. He has discipline and sticks to it.

The goal of a business is to show a profit.

A small used pickup will handle any chore that arises from a custom grazing business. Resist the urge to buy a fancy $40,000 dually 4x4 truck. You need to be lean and mean especially in your early business development years. Remember the old saying "No pain no gain."

If you can resist this new truck urge, then you're on your way to becoming a profitable grazier. Your new business venture does not need the strain of a new truck payment every month.

I see a lot of young guys 20-30 years old still living at home with their parents, but they have the latest fanciest 4x4 truck parked in the driveway. Most of the time they have a new ATV in the back of it.

Driving a new truck is a society thing. People think that they will be looked up to as being successful in life if they have a fancy truck. Nothing could be farther from the truth. Their new truck keeps them broke by consuming all their capital. That's why they're living at home with mama. The truck owns them. They don't own the truck.

I have a very successful elderly cattleman friend who drives an old two-wheel drive truck. Every morning he goes to get a cup of coffee at the local restaurant. All the locals will ask, "Old Timer why don't you park that old truck and buy a good new truck?"

He replies, "Why would I want to do something stupid like that for? There is nothing wrong with the truck I'm driving. It's just broke in." This elderly gentleman could buy a new truck every day of the year. It would never hurt his bank account. His discipline of not wasting money on new trucks every other year is one reason why he has accumulated a mass of wealth over the years.

Driving a new truck is a society thing.

I used to work with a fellow who would buy a new top of the line truck every year. He did this for the 21 years that I knew him. Once he bought the truck, he would buy every add-on that he could get his hands on to make it look even sharper. After he ran out of things to put on it, he would park it in his garage. Then he would go out and buy another older truck or car to drive to work.

I never could understand his logic of going to work everyday, just to make the payments, insurance, license, sales tax and the personal property tax on a new truck that was sitting in his garage at home. Yet every year he would howl about how the dealership was trying to steal his one-year-old truck at trade-in time.

This guy had a farm for awhile, but never seemed to have any money. He sold out and moved into town. The worst part of this story is when he reached retirement age, he told me that he had zero savings. He was worried sick about how he was going to meet his living expenses in retirement. He literally had spent every dollar he made on buying new trucks. He was a classic example of someone infected with the "New Truck Syndrome."

You may think that I'm being awfully nit-picky about something that doesn't seem to have any ties to a grazing operation. Maybe so, but if I can convince you to keep from purchasing your first new truck, you will have a 100% better chance of developing a successful grazing business.

You know what goes nice with a fancy new pickup? A fancy stock trailer, new horse trailer with fancy living quarters, a new ATV to put in the back of the truck, a new camper shell, a fifth wheel travel camper, a tandem axle utility trailer, a trailer full of jet skis, that new boat. The list goes on and on. It is a viscious cycle that will ruin your chance at ever attaining any kind of long lasting wealth from your grazing enterprise.

How are you going to build any equity or have any funds available to grow the business if it all goes to a truck payment? It takes a lot of pounds of stocker gain to pay off a new truck. What a waste of equity. You could have invested it in owning your own stocker herd. The owned stocker herd is going to show you a profit at the end of the year on cheap grazed grass gain. That new truck is just going to be tugging on your wallet every time you fire up the engine.

It takes a strong individual to resist the buying spree, especially when all your neighbors are rolling by you every day in their new shiny pickups and you're driving an older used truck. The truth is that a lot of these new truck owners would lose their trucks to the bank if they lost their job tomorrow.

Your insurance is tremendously higher and personal property taxes are higher with a new truck. Depreciation is a real killer on a new truck. The second you drive it out of the dealership its value drops three to five thousand dollars. When you finally do struggle through three to six years of paying for it, it is worn out and you feel like you need another one. Got to keep up with the Joneses, you know!

Your new business does not need the strain of a monthly new truck payment.

I had an accountant tell me, "Son you need to buy a new truck or stock trailer to show some farm expenses. You showed a profit this year on your farms."

I replied, "I thought that was the goal of a business to show a profit?" Needless to say the red-faced accountant changed the subject very quickly.

If you have to pay taxes at the end of the year, that is a good thing. Your business is correctly structured to make a profit. Pat yourself on the back. You hear ranchers brag that they have never paid taxes. This means their ranch is structured to fail. This is not something you should feel good bragging about! It is demoralizing to your soul to work all year long and show a loss every year. After several years of this, it's easy to adopt the attitude of "Why try? I will show a loss anyway." They better have a good off the farm job, or a good inheritance to support their love for the ranch or they won't be ranching long.

Chapter 15
The Economics of Leasing Pasture

The most costly ingredient of any livestock grazing business is the land. We have all heard the saying, "They are not making any more of it." This phrase has led many young ranchers into debt. They think that the price of the land they buy will always go up. I sure wouldn't want to bet the farm on it.

Remember the early eighties? Land prices plummeted. Farmers went bankrupt by the thousands. Farmers had bank farm loans against the supposed value of their land. Bankers were handing out farm loans like candy to babies.

Farmers were buying new equipment, new state-of-the-art buildings, and more overly priced land. All of this collateral was based on the present value of their land. Farms that were handed down from one generation to the next were now owned by the banks. There are still some farms around me owned by the banks, as a result of the land collapse of the 1980s. There is a chance that the land may go up, if it is located fairly close to an urban area. However, I am not interested in running a grazing operation next to a group of apartment complexes.

Allan Nation in one of his editorials really drove a point home in relation to owning land today. Most of us think that we can sell our farms when we get ready to retire and live out our golden years on the proceeds from the farm sale like our folks did in the past. Well guess what? The baby boomers are the buyers of the older generations' farms today. Who's going to buy our farms so that we can live out our retirement years from our farm proceeds? The baby boomers did not have the kids that their parents did.

Leasing land requires a change in your mindset.

This would lead one to believe that there will not be the wealth of buyers out there when we get ready to sell our land. A glutton of land and limited buyers means low prices for your farmland. It's the same situation that our grain crops are in today. It is hard for me to disregard this logical theory, which is another reason why leasing land is even more attractive to me than owning it.

Leasing land is a good way to make a profit every year. There is no worry of that dreadful land interest and principle payment that makes you a slave to it every year. I can remember the constant stress that I had when I bought my farm. What if I lose my job in town? What if I get sick? What if I have a drought? What if the bank raises interest rates? What if the price of my land plummets?

Guess what? It is a proven fact that if you can limit the stress in your life you will live longer. We have enough stress in our lives, without purposely adding to it by signing our life away to a piece of land by buying it.

Well I'm here to tell you leasing land is a lot less stressful than owning it. You don't have the worry of interest rates going up. You don't have a land interest payment staring you in the face anymore. 100 % of the lease payment is tax deductible. It is a business expense of your grazing enterprise. There are no land taxes, because you don't own the land.

It always burns me up when I receive the notice that my land is re-assessed. Government officials are never happy with their current level of your tax money, so they raise the value of your land to get more of your money. I know we have to pay taxes to maintain all public services. I just don't want to pay more than my fair share.

It seems like they know how to spend our money better than we do. My farm lays right on the boundary of two counties. I get assessed values and land tax bills from two separate counties. Two years ago, I got a notice that my land value had gone up five times over what I paid the previous year. I immediately called the county assessor and demanded an explanation as to why. He replied, "Well Mr. Judy, your land is in our classification of forty acres or less. We give all forty-acre or less land tracts in our county a classification of 'Recreational Property.'"

I informed Mr. Assessor that his supposed "Recreational Property" was part of a 200 acre working cattle farm. The assessor replied, "What do you want me to do about it? It has already got its classification."

I replied, "You messed up, you fix it, period."

The assessor regretfully took it back to court and got its classification changed back to agricultural land. I must have built two schools, twenty fire stations, fifty bridges and a hundred miles of blacktop with all the tax increases that they have levied against me in the last twenty-five years. Seriously though, leasing land does not require you to pay taxes on it. For what you can lease idle

> *Leasing land is a good way to make a profit every year.*

land for, it is financial suicide to buy land and try and pay for it by running livestock on it. If you buy the land, you had better have a very good off the farm job to make the land payments and have enough left over to pay for your living expenses.

Leasing land very simply means freeing up your equity so that it can be invested in growing your grazing enterprise, owning livestock, etc. instead of making land payments. Leasing land has allowed me to grow my custom grazing business from 40 head of stockers to over 1100 head in a period of three years. It gives you tremendous leverage when starting out as a young grazing business. It is scary the amount of idle land that is out there. All you have to do is look for it. I get cold chills when I look around the country and see all the idle land not being used for anything. The best part is, a lot of it is very economical to lease. There is a lot of work invested in getting the land set up, but the friendships formed, profit returns and personal satisfaction are all worth it.

Leasing land is a lot less stressful than owning it.

From a new fledgling business standpoint, it makes no sense to buy land when starting from ground zero. How many new businesses build a new building when first starting out? You simply cannot survive, because at the beginning of the business, it is not generating enough revenue to pay the bills.

I did buy a farm right at the start of my grazing operation. It was no picnic keeping the bank from taking it back though. It was very stressful and disappointing to work the entire year and give all your earnings back to the bank for your land loan interest payment. It seemed like the very minute the interest payment was made, time speeded up and it was time to make another interest payment before I knew it. There was never any money left to put a sizable payment down on the principle of the land loan.

The only reason that I was able to hold on to my farm and house was that I switched to leasing farms to grow my business. I was simply able to grow my grazing business because I found a huge leverage tool (leasing idle land) and took advantage of it.

Fellow graziers, there are thousands of acres of idle land out there to manage. You may have to be innovative to make them into profitable grazing systems. Develop water systems, install some fence, broadcast some seed, that's the most rewarding part of leasing, developing the grazing system. The more farms you develop, the stronger your confidence grows.

It is impossible in my area to buy a farm at the going rate of $1200 - $1500 per acre and expect to pay for it by grazing cattle. It just plain isn't going to happen folks. Maybe if you could find land for $300-$400 per acre, you might pay it off in your lifetime, although it will be a long struggle if you're starting from scratch. There is no such land around me to buy for that price anyway.

A young couple I know bought a 100-acre grass farm for $1200 per acre, which gives the farm a price tag of $120,000. He doesn't know beans about grass or managing it. He firmly believes that the more hay he puts up from his farm, the more money he is going to make. His father-in-law loaned them the money for the down payment on the farm. This farm has only 60 acres of open ground on it for grazing. The rest is brush and creeks.

Leasing land means freeing up your equity.

He figured he could run 30 cows year round on his sixty acres of grass and also have enough grass to cut all the hay he would need to winter them. He is also getting a loan to buy 30 bred cows (top of the cattle cycle no less) to graze on his grass. Take 30 cows x $1200 each = $36,000 invested in his cows.

He calculated selling 30 calves each fall after weaning them. He was going to creep feed them pellets so that they would weigh 600 lbs as a weaned calf. He also figured selling them for a $1.00 a lb at the sale barn. Well the way he had it figured, he would have $18,000 worth of calf crop each year generated from his 30-cow herd.

About this time I felt like waving some smelling salts in front of his nose to wake him up out of his dream. I didn't want to hurt his feelings, so I shut up for the time being. I hope he gets a live calf from all of them, although that's not likely. He has already bought a 60 hp tractor, disc, plow, bush hog, baler, rake, mower and grain drill. All the equipment was bought for a total of $30,000. He has a separate equipment loan.

Both he and his wife work in a factory making $9.00 per hour. His wife wants to quit work to stay at home and take care of their new baby. Sadly, I don't think there is much hope of that happening if they don't want to lose the farm. I'm praying that neither one of them are laid off from their factory jobs.

All of his loans total $186,000 for land, cattle and machinery. Let's assume the 100-year-old cattle cycle of going up and going down has been broken. Cattle prices will never go down again. Calves will always be a dollar or more a pound. Here are what his owned farm numbers will look like, based on his assumptions:

1. $186,000 farm loan x eight percent interest = $14,880 due yearly. (No principle included)
2. He will sell 30 - 600lb calves for $18,000.
3. 30 cows x $329 (1 year of cow care) = $9870.
4. We will assume no machinery costs. (Everything runs on air and sun.)
5. We will also assume no depreciation on cows, machinery or buildings. (His cows will live forever.)
6. We will also assume 100% calf crop because he figured it so.
7. Heck we won't even charge the calves for their creep feed.

Let's do the math now:

1. Total farm bills including cost of maintaining cows and land interest = $24,750.
2. Total gross income generated from owned farm = $18,000.
3. Farm loss = $6750.

Not a pretty picture by any means.

I got to feeling sorry for what this young couple was getting ready to face. I stopped by his work one day and talked with him on his lunch hour about Management-intensive Grazing and other related topics.

I might as well have been talking to a wall. All he wanted to talk about was his recently purchased machinery and how much hay he was going to be able to bale that summer.

I didn't give up though, I loaned him several copies of *the Stockman Grass Farmer* to read and recommended that he subscribe to it. A week later I saw him and asked him what he thought of the magazine.

He replied, "Heck all they talked about in that magazine was grazing grass. That's boring stuff. There was not one article on tractors or balers, which are your more important stuff."

I asked him for my magazines back and wished him good luck. Some people are fixed in a certain mindset and they will not change. I'm not a betting man, but this young couple is almost a 100% sure bet for losing their farm. They have a better chance of getting hit by lightning, than they have of ever paying off their farm. It doesn't need to be this way at all.

Learn to love doing things differently from everybody else.

Leasing land requires a change in your mindset. Human beings are resistant to change because it makes us feel uncomfortable to be away from familiar surroundings. We like to follow the flow. It's easy to follow what somebody else is doing. It has already been proven that it works. What's rewarding about walking down a beaten path? You need to learn to love unfamiliar surroundings, doing things different than anybody else has ever done them. This is what keeps daily life so intriguing and challenging. A person needs to be challenged in life to grow. Remember, the brain is like a muscle, the more you use it, the bigger it grows.

Chapter 16
Low Risk Custom Grazing

Custom grazing is a great way to learn the grazing business with no risk and little equity. I cannot think of a better way for a grazier to get his feet wet. I would suggest starting out custom grazing with a group of dry cows and learn the ropes of MiG. Dry cows are a lot more forgiving than stockers while you're learning to keep your grass vegetative. Dry cows do not have the nutrient requirements of a young stocker calf. You will make some mistakes, but none of them will be career ending though.

You will learn more from the mistakes you make than the things you do right. Remember you are not perfect, so look at mistakes as learning exercises that you will not soon forget. There are thousands of calves out there available for grazing. You can get your share of them if you concentrate on growing grass on leased land. It is extremely important to manage the calves as if you owned them. You will get repeat customers if you can hold down death loss and put on lots of grass gain.

I first got into custom grazing when a friend asked me if

I would be interested in pasturing some yearlings for him. He knew that I was trying to survive the lengthy divorce that I was involved in. Legal proceedings had consumed all my savings. All my livestock were sold to pay legal fees. At the time things looked pretty grim, but I had 50 acres of improved pasture set up in 12 four-acre paddocks. I had one year of MiG under my belt and was really sold on the forage improvements and gains that I had seen in one year with my own stockers.

> *Start custom grazing with a group of dry cows to learn MiG.*

I jumped at the opportunity to graze this group of mixed steers and heifers. I helped the stocker owner haul them over to my farm. We weighed the truck and trailer empty first, then weighed each load of calves before unloading them on my farm.

We each had a copy of the total weight of the 49 stockers. These calves were out of a good bunch of beef cows. The owner really had been working on improving his cowherd.

I immediately received a summons from the sheriff that I was to appear in court for a special hearing. My ex's lawyer heard that I had calves on the farm, and he assumed that I had bought the calves with a hidden pot of money! The judge was not very happy with the ex's lawyer for wasting valuable court time after I proved that I did not own one single head.

I moved the calves twice a day in the month of April. Through May, the stockers were moved once a day. Through June the stockers were moved every two days. All paddocks were fed with pond water. July 15th the calves were loaded, weighed and sold. They averaged two-lbs of gain per day on that bunch of 49 stockers. I was ecstatic. Normally around my area the average was .75-1 lb/day gain.

Shortly after loading the stockers, drought hit, corn prices went up and cattle prices went down. I was unaffected by both because my farm was in a rest mode waiting on the fall growth. The drop in cattle prices did not affect me because I didn't own any livestock.

Custom grazing gives you tremendous flexibility when managing your grass. You can decide how many you want in the spring at turnout and get them lined up for the grass explosion. There are no trips to the banker to ask for a loan. Just get busy lining up calves.

Where Do You Find the Dry Cows?

Place ads in all the local auction barns, feed stores, vet clinics, newspapers. Keep the ad short with easy to read words. It looks more professional if it is typed in large black letters for posting on bulletin boards. Ask the local sale barn owners if they would be interested in providing you some calves to graze. Ask your neighbors who have cattle if they would be interested in letting you graze their cattle on your leased land. Explain to them how you will move the cattle from one pasture to the next to ensure they always have fresh grass and how that allows the other pastures to rest.

I custom graze both steers and heifers on my leased farms. I want to remain flexible for my stocker owners. You have to learn to work with your stocker owners and the cattle price cycle. Sometimes steers get too expensive to buy so the cattle owners switch to heifers, which are always cheaper. When grazing heifers you have the worry of a neighbor's bull getting in with them and breeding them. A good hi tensile perimeter fence will solve most nuisance bull problems. Heifers will not gain quite as well as steers in most cases. Usually you'll average 8 to 12 dollars more per

> *You will learn more from the mistakes you make than from the things you do right.*

steer of custom gain than from grazing a heifer. But sometimes steers are not available so remain flexible and be willing to graze heifers. It's better to have a field of heifers than a pasture full of seed heads from not having enough grazing pressure when the spring rush hits.

Custom grazing gives you tremendous flexibility when managing your grass.

Cow owners like to hear that their cattle are checked at least every two days. When you get your first set of cows, manage them like your life depended on it. If you take good care of these cows, the word will get around that you are a good stockman.

When the owner comes to pick up his cows and you have managed to put some nice fat cover on his cows over the grazing season, this makes your grazing management shine. These cow owners may become some of your future stocker calf owners that you custom graze for. Remember, your best advertisement is a satisfied customer.

I like to have some heavy calves at early turnout simply because they can remove a lot of forage. A 600 - 700 pound calf will remove half of the plant, where a 400-pound calf takes the top two to three inches. You need to hog off that plant in April-May with a heavy stocking density or you will have a pasture full of worthless seed heads.

You don't want to be sitting on a tractor seat in May mowing off seed heads to keep your grass vegetative. This is exactly what will happen if you are not stocked heavily when the spring grass rush hits.

I like several dozen 700-pound calves with every group of 400-pound stockers that I turn out in the spring. These 700-pound calves really mow the forage down and pay you for it while they're doing it, unlike sitting on the tractor seat where it is costing you money. As a grazier, there is nothing more depressing than mowing off seed heads, simply because it means you missed an opportunity to put some cheap grass gain on a growing calf.

As a custom grazier I assume no responsibility for death or loss of calves. The stocker owner has a liability policy in effect on the calves when they are placed on the leased farms. I give their insurance company a legal description of the farm. You must have a written contract for each bunch of calves you graze stating all the rules that the stocker owner and the grazier agree on. This prevents confusion later in the year when questions arise during the contract period. The lease also needs your signature and the owner's signature.

I write the initial contract and then ask the stocker owner to read over it and mark up any concerns he may have about it. I cover everything in detail, who is responsible for each item, the cost of gain, doctoring, loading date, etc. Always leave yourself an out in case of a drought. This is a huge advantage with Management-intensive Grazing. You can count the days of feed left in front of you.

I will give the stocker owner two to three weeks notice in the case of drought. This usually gives him time to line up some place to go with them or make arrangements to sell them. As bad as droughts are, they are a lot easier to stomach and survive if you are de-stocked and letting your pastures rest. This is another huge advantage of custom grazing.

You have to learn to work with your stocker owners and the cattle price cycle.

Listed on the next page is what the actual contract looks like. I numbered each sentence in the contract so that I could explain the importance of each. At the end of the contract is a detailed explanation for each sentence.

2002 Grass Grazing Contract

1. Greg Judy agrees to custom graze 100 steers from April 1st to August 1st for the owner John Doe.
2. Cost of gain will be .32 cents per pound of gain.
3. Greg Judy is responsible for all water, electric, grass, salt, and labor for doctoring and moving the calves.
4. Calves will be rotated every two days or as grass conditions warrant.
5. John Doe takes full risk of death loss of any steers.
6. For any calf that dies, we will take the average spring weight and subtract from the gross spring weight ticket.
7. In the case of a drought, Greg Judy will give John Doe three weeks prior notice when he wants the calves loaded.
8. John Doe will give Greg Judy a gross weight ticket of the 100 steers at the spring turnout.
9. John Doe will give Greg Judy a gross weight ticket in the fall after they are loaded.
10. John Doe will pay 0.16 cents a day per head, which is due at the end of each month.
11. The remainder of stocker gain will be due when the final weights are received from the August 1st gross weight tickets.
12. John Doe agrees to pay for all hauling of steers.
13. John Doe agrees to have a liability policy in effect on all steers when they are placed on Greg Judy's farm.
14. Grazier Signature: Date:
15. Cattle Owner Signature: Date:

Sentence 1: This sentence lists the cattle owner's name, the grazier's name, the number of steers to be grazed and the grazing season's starting and ending dates. It is very important to have the dates on the contract so there is no mix up between the two parties as to when the grazing season starts and ends.

Sentence 2: This sentence tells the cattle owner the price

of gain that the grazier is charging for each pound that is put on the steers.

Sentence 3: The grazier is listing all the items and services he will provide in the grazing contract.

Sentence 4: The grazier is describing the grazing rotation the steers will be subjected to.

Sentence 5: The cattle owner assumes all risk of any death loss.

Sentence 6: This sentence explains how death loss weight will be subtracted from the gross spring weight.

Sentence 7: The grazier agrees to give the cattle owner three weeks notice to remove the steers in case of a drought.

Sentence 8: The cattle owner agrees to give the grazier a gross weight ticket at the time of spring turnout.

Sentence 9: The cattle owner agrees to give the grazier a gross weight ticket at the time of fall loading.

Sentence 10: The cattle owner agrees to pay 0.16 cents per day per steer, which is payable at the end of each month.

Sentence 11: This sentence explains that the balance of the steer gain will be due when the steers are loaded and weighed in the fall.

Sentence 12: The cattle owner agrees to pay for all hauling of steers.

Sentence 13: The cattle owner agrees to have a liability policy on all steers. This protects you from being sued if a steer would get out and get hit by a car.

Sentences 14 & 15: The grazier and landowner signatures.

Custom grazing allows a grazier to build equity while the cattle cycle is at the peak. There is no fun buying a steer in the spring for $1.20 a pound, then having the price drop to 75 cents a pound in the fall when you sell him. This takes all the fun out of grazing grass very quickly.

The cattle owner has the fixed costs of interest, hauling,

medicine, minerals, death loss, and the possible negative margin when selling a heavier calf. To cover the costs of the cattle owner purchasing and hauling the calves, you need to put a minimum of 200 pounds of custom gain on the stockers, preferably 250 to 300 pounds. For the cattle owner and custom grazier partnership to work, you both have to make money or the partnership will not last. Lots of cheap grass gain ensures a good profit for both the cattle owner and yourself.

When it comes to negotiating a price to charge for grazing calves there are several things to consider. First make a list of the services and materials that you will provide in the proposed grazing contract.

The more items you provide, the more you should charge to cover your expenses and leave yourself a decent profit.

Most custom grazing fees may start at 30 cents and run up to 40 cents per pound of gain all depending on the individual contract. Some grazing contracts may be negotiated for a certain cost per day for each calf. This type of contract is the safest guaranteed income for the grazier. However it does not reward a grazier if he is capable of putting on a high daily gain.

Let's look at an example of getting paid 32 cents per day for every day the calf is on our farm.

1. 160 day grazing season x 32 cents per day per calf = $51.20 per calf
2. 160 days grazing x 1.5 lb gain per day = 240 pounds gain per calf.
3. 240 lbs x 32 cents (custom gain charge per pound) = $76.80 per calf
4. $76.80 minus $51.20 = $25.60

Results: The custom grazier who was getting paid by the amount of gain that he put on the calves grossed $25.60 more per calf than the grazier who was getting paid a flat fee of 32 cents per day. This is a pretty nice incentive to learn to be a proficient grazier.

A good strategy would be to contract calves on a flat fee cost per day per calf until you become an expert grazier. Once you become proficient at putting on high daily gains, then charge by the pound. This way you are reaping the rewards of your grazing expertise that you have developed over the years. Notice I said years. You will not become an expert stocker grazier overnight. It takes years of learning to become an expert grazier. I'm not there yet, but I have made some huge strides and thrive on the challenge of getting better every year.

Once you get a load of calves, keep the cattle owner updated on their progress every couple of weeks. If you have to doctor a calf, inform the stocker owner that you did so. This alerts the cattle owner to the problem of calf sickness. The more open channels you can keep between yourself and the cattle owner, the more confidence and trust you will build.

Cattle owners appreciate you giving them an update on their investment. It shows them that you are a conscientious grazier and are truly concerned about the welfare of their calves. Remember a person is only as good as his word. Be honest with your cattle owners and you will be rewarded for it.

Chapter 17
Custom Grazing Through Winter

For winter grazing calves, you need to have a fixed yardage cost per day on each calf in addition to the standard grazing contract. What I mean by this is that let's say you have a grazing contract on a group of calves where you are getting paid by the pounds of gain you put on. You need to be paid additional yardage per calf per day for managing these calves in winter weather.

I usually start charging yardage around the middle of December in my area. This is when the calves take a lot more labor to care for just to meet their daily needs. The water needs to be looked after daily to make sure it is not frozen. You may have to feed some hay when the snow gets too deep. You have more mud and ice to deal with. The list goes on and on.

You are going to be exposed to some tougher environmental conditions caring for the calves in the harsh winter weather. It's that simple. There will be snow, ice covered ground, cold wind, possibly ice to chop. Believe me, it is not the same as summer grazing on green grass. You need to be paid for the extra labor it takes to take care of the calves.

The cattle owner cannot expect to have his calves wintered for nothing. If he had them in a feedlot, they would charge him yardage also, plus a feed bill for every day they are in the feedlot. Depending on where you live and the severity of your winters, this should dictate what you charge for yardage. For the calves that I winter, I target them to gain around 1/2 pound gain per day. I don't want to buy expensive supplements to make the calves gain one to two pounds a day during the winter. This would be economic suicide.

Winter is as long as you make it.

I guarantee that you will lose money on every calf if you follow this strategy. The local feed store will give you a nice jacket and love you to death for being a good feed patron. Your billfold will also be easier to pack in your back pocket, because it will not have anything in it.

The calves that I winter are building a lot of compensatory gain for the spring rush of cheap green grass. It still absolutely amazes me how fast a wintered steer puts on weight in the spring when he gets all the tender new grass he wants every day. You don't want to have the calf real fleshy when you are coming out of winter. A fleshy calf does not have the available room on his body needed to store cheap grass gain. I don't care how cheap grain gets, it will never be able to compete with cheap leased grass gain, period.

Winter grazing does provide some nice cash flow through the winter. It also provides you with some nice calves with huge appetites to take advantage of the spring rush of grass. It truly is a wonderful feeling to have a large trained group of hungry mowing machines when the grass is trying to get ahead of you. A developing seed head waits on no one.

An advantage of leasing multiple farms is that I can stockpile a lot of forage on several of the farms strictly for my winter grazing groups. These targeted wintering farms have the stockers removed from them at the end of July. No other stockers are grazed on these farms until November. By this time there is a beautiful sward of stockpiled forage to winter a group of stockers on. I must emphasize that this stockpile is 70 percent fescue, 30 percent red clover. I have enough legumes in the pastures that no commercial nitrogen is applied for fall fescue growth.

Wintering a group of stockers is a great way to build up the fertility of nutrient-starved pastures.

A lot of people in my area cuss fescue, but I love it to death. There is no other forage that you can abuse as badly as fescue and have it still maintain a thriving grass stand. It forms such a dense sod that it will hold up livestock in the most wet weather that mother nature can throw at you. Fescue makes a wonderful forage as long as you keep it vegetative and have some legumes mixed in with it.

Fall-grown fescue does not have the large amounts of toxin in the plant that spring grown fescue has. The spring-grown fescue has seed heads and stems, which is where the toxic endophyte is stored. All the leased farms that I have are mostly Kentucky 31 fescue base, with 30 to 40 percent red clover mixed in.

I use two separate methods for feeding out stockpiled grass in the winter. The most efficient method of rationing out your precious grass stockpile is by strip grazing. I give the group of calves just enough grass for them to completely clean it up in two days. Then the poly wire is moved forward for another two days worth of grass. You will develop an eye for how much stockpile to put out with each temporary paddock as you go along. Make them clean it up completely before exposing any fresh stockpile.

The stockpiled fescue in my area still has green leaves deep down in the lower section of the thatch in February. The calves will walk away from hay every day of the week to graze good quality stockpiled forage. You need to adopt a different mind set when it comes to rationing out your winter stockpiled grass. Give the calves enough grass to replace the hay that they would normally be eating. If you give them more grass than this they will eat it, but remember, you're just giving them enough to get by on until the spring grass gets here.

Treat your stockpile like it is a huge pot of gold, because it truly is. You want to start strip grazing the calves in the area of the water source first. By doing this the water source is always available to the calves as you move your wire forward. Make sure you have a minimum of 4000 volts in the poly wire.

I cannot stress this enough, a hot fence is a must with strip grazing. If the fence loses voltage for whatever reason, the calves will discover it in less than two hours.

I had a herd of calves that was very well trained to electric fence. These calves were never out, always right where you put them. They knew that the white poly wire was the devil himself. One night the fence got grounded out when something dragged a piece of old barb wire up over the perimeter hi-tensile fence. Needless to say the trained group of calves had helped themselves to a paddock change in the middle of the night. Keep it hot and the calves will be in the right spot.

A hot fence is a must with strip grazing.

Always use white poly wire, they can see it a lot better and will not be as likely to run through it. Poly tape will work, but it is a lot more burdensome to work with and more expensive than poly wire. The tape will also go down in any ice storm that you may have.

I use tread-in posts sometimes, but I prefer to use 1/2

inch fiberglass posts spaced 60 to 90 feet apart. The fiberglass posts make a stiffer fence and are more deer proof than tread-in posts. The 1/2 inch posts are cheaper in some fencing outlets than tread-in posts. These posts are only driven in the ground five inches. A cotter key is used to hold the poly wire in place.

Adding free fertility to poor pastures is a main reason for wintering calves.

These posts drive in the ground great until you have a week of severe cold weather. Then you simply can not drive them in the frozen ground. This is when you need a three foot steel, 1/2 inch rod that is sharpened on one end. Take a hand post driver and drive the steel rod sharpened end five inches into the ground. Remove the steel rod and slide a 1/2 inch fiberglass fence post down in the five-inch-deep hole that you just made with the steel sharpened rod. Lightly tap the fiberglass post down with the hand held post driver to tighten the post in the hole. Ree rod steel posts will work with a good plastic insulator attached to them. It takes a little more time when the ground is frozen to set up your temporary strip paddocks, but you are saving some serious money every day that you do not have to feed hay.

I will lay out three to four temporary paddock strips to be grazed ahead of time. By doing this I have a week's worth of feed ready for the calves by simply removing the poly wire and exposing a fresh two day stockpile of forage every other day. I feel a little bit guilty when I go out and feed 100 calves in five minutes by simply dropping a wire, while everyone else is using their big tractors and trucks to feed hay every night. Leasing idle land gives you tremendous flexibility to stockpile valuable forage to ration out through the winter. Without the leased land I could not economically winter anything.

The second method that I use sometimes is rotating the

calves daily through the stockpiled paddocks. This is not nearly as effective as strip grazing, but it is an enormous time saver. There is no wire to move each time you are ready to change paddocks. Just open the gate and let them have the entire paddock. By moving them often, they do not seem to trample the stockpile down as bad.

Don't get concerned if they leave lots of forage with the first grazing rotation. They always leave some, because once they walk on it, they don't eat it as well.

On the second rotation when I come back to this paddock usually a rain has cleaned off the grass and removed the smell of hoof traffic. When you turn them in to this freshened up paddock they go after it like gang busters. They eat it just as well as a fresh ungrazed paddock, assuming that the paddock has been rained on.

Depending on the amount of your beginning stockpile you may get as many as four rotations through each paddock. The key is to not let them graze on a paddock any more than one day or they will trample the grass down so badly that they will have a hard time getting their tongue around it the next grazing rotation.

Another advantage of moving them daily is the manure distribution is better because they do not have a chance to set up a favorite lounging spot. Another reward that I have noticed with continuous winter grazing is the calves do not use the shade trees nearly as much. I find 90 percent of the manure is deposited right out in the middle of the stockpiled field that they are grazing that particular day. The field usually has exposure to full sun, when they get full from grazing the stockpile, they lay down and chew their cud where they can soak up the sun.

Treat your stockpiled grass like a huge pot of gold.

Guess what they do when they get up? Manure, urinate then go back and start eating again. It's a pretty good life, and beats standing around a bale ring in mud fighting other calves to get at a bite of hay.

Strip grazing is the most efficient method of rationing stockpiled grass.

Another added advantage from winter grazing is the sprout removal in the paddocks. When you strip graze correctly they literally eat everything except for the multiflora rose bushes. But they do trim up the rose bushes nicely, which makes it easy to get at them with an ax.

A word of advice on the subject of hay. Never own any hay equipment. You cannot afford to have that equipment sitting around rusting and depreciating with each year that goes by. I am convinced that some people really love to cut, rake and bale hay. It makes them happy as long as they are in the hay field making hay. So by all means I am going to do my part to let them be happy. I buy all my hay and feed it on the leased farms.

Never sell any hay off any of your land. By selling the hay, you are selling valuable fertility that will have to be replaced or your pastures will suffer for it.

I calculate ahead of time, 1/3 bale (300 lb) of hay for the entire winter for each 500 pound calf. This time period covers 150 days, 120 days are on grazing stockpiled fall forage and 30 days are on hay. You need an emergency supply of hay in case the weather turns sour. It is sometimes impossible to graze when you have two to four inches of ice over all the stockpile. Not only is it hard for the calves to get at the stockpile, but they tromp it down much worse when it is covered with ice. This is the time that it is better to feed your emergency hay reserve. I always figure 30 days for feeding hay in my area. It may require more hay in other areas of the country that have more snow and severe winters.

I did hear an interesting quote several years back from a native New Zealander who was talking at a conference. A fellow in the crowd asked the New Zealander, "Well how do you get your livestock through the winter on just grass and no supplements?"

He responded, "We don't have the grain. That is why we depend solely on our grass." He summed it up by saying, "Winter is as long as you make it."

The fellow in the crowd who asked the question just let out a big scoff and replied, "Well it will not work here in America. My cattle would starve if I did not feed them some grain and a steady diet of good hay through the winter."

Basically what the New Zealander was telling the crowd of cattle producers was that we in America are dependent on the glutton of grain and that is why we do not know how to grow and utilize good pasture. I think about the New Zealander's quote every time I move my poly wire. Winter is as long as you want to make it!

I purchase my 30 day emergency hay in October and set the bales out on the individual paddocks to be unrolled in the winter. The bales are set on 25-foot centers against paddock division fences in the poorest areas of the farm. I designed a bale unroller that I pull behind my small Toyota pickup. It works wonderfully to distribute the bale out across the paddock and the best part is you can do it from the warm cab of your truck.

By leasing multiple farms you can stockpile a lot of forage for winter grazing.

There's no tractor to worry with starting in cold weather and no ruts to level out in the spring, just beautiful windrows of manure across the paddocks. The calves will always lay down on the wasted hay because it is dry and keeps their body insulated from the cold ground. It does not bother me to see a little wasted hay that the calves manured on. This is valuable organic material for the soil.

In winter target calves to gain 1/2 lb per day.

The unroller was made from a set of spindles that I had on an old trailer. I had a welder friend build it for me for 150 dollars. Basically the way it works is you back up to the big round bale with the unroller. It has two spikes that are mounted in a steel hub above each tire spindle, which follows the bale to the ground as it is unwound. The spikes are simply driven into the center of the bale to a depth of two feet before you start to unroll it. It also has a two ton ratchet hoist that makes it possible to move the bale anywhere you want to before you unroll it.

This is a very handy feature because sometimes you may have an isolated spot on the pasture that really needs some added fertility. I just ratchet the bale a couple inches off the ground to travel over to the particular poor spot and unroll it. The next morning the poor spot is a solid windrow of wonderful manure. The cattle spread it for me. No tractor or manure spreader are needed.

Wintering a group of stockers is a great way to build up the fertility of nutrient-starved pastures. By picking out the poor spots, you are putting the manure right where it is needed most. This ranks as one of the top reasons I winter calves, just for the fertility advantages that you can capitalize on.

I have some very strong dense grass stands in areas that would only grow broomsedge before I unrolled hay bales on them in the dead of winter. It is a wonderful feeling to see an old previous dead area of the pasture that is now thriving with green succulent grass. By unrolling the purchased bales of hay, I am adding free fertility that normally would have to be purchased, plus the calves are getting a meal to boot. I know of no other way that is as economical as unrolling hay to help build up your pastures' soils. The calves do all the work. You're just giving them the tools to complete the job.

The economics are great with a pickup-powered home-made bale unroller. First of all you don't need a tractor or a large truck to feed hay. Secondly you can feed 100 head of calves with one big round bale of hay per day. They all have access to it because it is strung out across your pasture for several hundred feet. With bale rings, only 25 calves can eat from one big round bale. For 100 calves you would need four bale rings and have to set out a minimum of four bales every two days.

Another bad effect of feeding hay in rings is the pugged area around each bale ring that takes years to grow back in grass. Feeding hay in bale rings also deposits all the manure, urine and wasted hay in one twenty foot circle. This 20-foot ring of nutrients does not do much for your nutrient-starved leased pasture land. The bales are fenced in with white poly wire. When it comes time to feed one, simply pull back the poly wire and unroll the exposed bale. Then fasten the poly wire back, protecting the remaining bales.

I put five to six bales in every paddock. That way when it gets a lot of hoof traffic on one bale feeding area, I will move to another paddock and feed a bale there. That way you are never tearing up your sod. Even in severe wet periods, it really limits pugging if you will move the calves to a fresh paddock every couple days. Another advantage of wintering groups of calves on your pastures is they clean off all the duff, which makes a perfect seed bed for frost seeding red clover in late winter.

Where you live and the severity of your winters should dictage what to charge for yardage.

You will never have a problem getting legumes established in a winter strip-grazed pasture if it has adequate fertility. You have simply removed all above ground competition. The baby legume seedling can really get a good foothold for its life cycle before the fescue has a chance to shade it out. There is usually a little bit of dirt showing in areas from severe hoof traffic, which even ensures a better legume seed bed.

The grass also gets off to a slower start in the spring because of the severe sward removal through the winter. This weakens the grass stand enough that the young clover seedlings have excellent sprouting and growing conditions. Remember to use a good inoculent on those clovers so that the legumes can maximize nitrogen fixation from the air and store it in their roots for future forage growth.

Chapter 18
Handling New Arrivals Gently

Spring turnout calves should have their second round of shots and be wormed before being placed on pasture. The first pasture they are turned in on should have a minimum of two hot wires, preferably three to four wires.

Your fence should be energized to a minimum of 4000 volts. New arrivals need this much pain potential to break them in right from the start. The quicker they receive their first unpleasant experience with a truly hot fence, the less trouble you will have with them getting out.

This pasture should not be any larger than two to three acres and have 3" to 6" of tender grass available. Good clean pasture of tender grass is the best medicine you can give a newly arrived calf. A calf that is eating is one that will go on and start putting weight on and not get sick.

Water should be easily accessible to all calves. I prefer a water tank that is against one of the perimeter fences. By having the water tank against the perimeter fence, the calves will actually walk right into it while they are finding the boundary lines. It is very important to get some water in the calves immediately because after being hauled they are usually severely dehydrated. Watch the calves when they are turned loose in the pasture. Make sure they all drink.

Just walk them up to the tank and hold pressure on them until they drink. Sometimes I will have a garden hose splashing water in the tank. This way the calves that are not used to drinking out of a water tank can actually hear the water running. Calves will always go to the sound of running water.

Good clean pasture of tender grass is the best medicine for newly arrived calves.

I run a 50 foot piece of white poly tape perpendicular to the perimeter fence. They will always go up and sniff it out of curiosity, and it is very visible. The calves get a quick meaningful lesson on electricity.

I take a five-gallon bucket and go out and sit with the new arrivals. I believe it helps take the stress off of them by just simply being there and letting them know that you're not going to hurt them.

To stop fence walking, you simply move slowly in front of the lead fence walker and sit down on your bucket. Do not get so close to them that you're in their flight zone, because they will start running. The lead fence walker will stop walking to concentrate on what you're up to, which brings the whole herd to a stop.

Remember they are herd animals. If one starts walking the fence, the whole herd will, and you will end up with a health wreck. They may go around you and continue walking. Simply pick up your bucket and assume your position in front of them again. I have had the whole herd just come out and literally lay down beside me because they are tired from being handled and hauled. They just want to rest and I'm giving them the opportunity to do so. It is a nice feeling to have the whole herd lay down around you so they can rest. You must have 100% quiet, no noise or sudden movements in this whole practice. In time your calves will be happy to see you. They know that you are going to move them to a new tender fresh pasture.

I highly recommend reading some of Bud Williams' cattle handling techniques or going to one of his schools that he holds around the country. He is the best animal handler that I have ever seen. He takes the stress off new calves by teaching them to handle pressure. Bud has a video showing exactly how he teaches calves to handle pressure. Bud has very little death loss or sickness, because he gets the stress off the new calves so quickly they don't get sick.

I have a lead steer that I constantly praise. I would hate to think what I would do without him. I named him EARS, because he is a Brahma and Charolais cross. He has these big floppy ears. My wife wanted to call him Dumbo after the Disney elephant character. I told my wife there was no way I was going to call such a smart steer Dumbo. What a magnificent animal he is. When I holler for Ears, he lets out this ear shattering bellow that can be heard for a mile. I love the sound of it. He will follow me absolutely anywhere I go with a bucket even though it doesn't have any feed in it. I will admit he is happier when the bucket has a little feed in it though.

I turn Ears in with each new group of arrivals and guess who they adopt as daddy? It sure helps remove the stress on new calves by having an older steer or cow with the young calves. It has a calming effect on them to be around an older animal. The sooner you can get the stress off them, the less chance they have of getting sick. Stress causes sickness, plain and simple.

Water should be easily accessible to all calves.

With Ears, I can literally on day one start moving new calves on rotation through the series of paddocks. When you have a whole series of paddocks that need grazing and on day one can start rotating them, this literally makes Ears priceless. Before I had Ears, a new group of calves might take several weeks to really get comfortable with me and move to the next paddock when I called to them.

To show you a financial example of what a lead steer means in dollars to your stocker operation let's look at the numbers. In Example 1 we will look at training new calves to move from paddock to paddock without a lead steer:

Example 1 (No Lead Steer)

Let's assume we get 200 new calves ready to graze. A conservative estimate would be two to three weeks before every single one of the calves was trained to move when you called them. On the first day the pasture has had the top 2-3 inches of candy grass removed. On day two the calves look at you when you call to them and keep their distance. You may get a handfull to go through into the next paddock.

Now you have calves in two paddocks. Your grazing machine is getting weaker because the calves are getting spread out over more area. They are bawling because they see their buddies several fences over and are not sure how to get to them. You need to keep them together, to make it easier to check and move when they are one large herd, and the calves stay calmer in a large herd. These calves are not gaining at all. They are stressed.

If you take the total 21 day learning period for getting the calves to move as one large herd, they would average between .5 to 1 lb a day gain. Of course this may vary on the type of forage they are on. But the bottom line is that the calves are not on quality forage every day, and they are stressed because they are split in small groups, and your pastures are getting more mature as every day goes by. Take 21 days x 0.8

lbs gain per day (average) = 16.80 total pounds of gain per calf for the 21 day period.

Take 200 calves x 16.80 lbs = 3360 lbs of total gain.

Let's assume we are charging 0.32 cents per pound of gain on each calf. Take 0.32 cents x 3360 lbs = $1075 worth of custom gain.

Now let's take a look at Example 2:

Example 2 (Using a Tame Lead Steer)

Day one you walk out in the middle of the pasture with the 200 new calves looking at you and you sit down on your bucket. Your lead steer, Ears, immediately spots you, lets out a bellow and comes running to get a mouthfull of grain. Other new arrivals follow to see what's up. They watch Ears hogging down the grain in the bucket while you scratch his head. You get up slowly to move. Ears follows, bringing the whole herd with him.

You quietly leave the paddock, concentrating on not getting too close to a steer's flight zone. Once you leave the pasture they go back to grazing.

Day 2 you will walk over to the gate and call for Ears. He comes charging across the paddock and you go through the gate after the whole herd has accumulated at the gate entrance. Do not go through the gate until you have the whole herd, or you will leave some that will not come through because there is nothing to follow.

These calves are placed on succulent pasture again on day 2. Heads go down and they go to grazing, topping the plants, taking just the best. This group of 200 calves would average gaining between 1.5 to 2.5 lbs a day. Let's take 21 days x 2 lbs gain per day = 42 lbs total pounds of gain per calf for the 21 day period.

Take 200 calves x 42 lbs = 8400 lbs.

Take 0.32 cents x 8400 pounds = $2688 worth of

custom gain for having a lead steer and moving through paddocks from day two on. That is a $1613 increase in profit by having a lead steer. This is just on one group of calves.

I had one wild bunch of calves that would literally hide in the woods when they saw me coming. I would leave the gate down with a bucket of grain by the entrance and it would still take them a week to get the bravery up to go through the gap.

If you have farms that are split by roads a lead steer is priceless.

Consequently the calves gained very poorly, 0.8 lbs per day. The grass ahead of the wild bunch was getting mature and the grass in the paddock that they were in was overgrazed.

If I had had Ears, he would have calmed that bunch down in two to three days. I would go out and sit with the wild calves and they would not allow me to get within a 100 yds of them. I tried bucket feeding them, but I could not get close enough to even let them know I had some feed. This was my second year of custom grazing and cattle prices were high, I took what calves I could get my hands on.

I will walk out in the middle of the pasture and sit on my bucket and Ears will bring the whole herd with him to watch him eat some grain out of my hand. Shortly there are other calves wanting some of the grain that Ears is chomping on. Ears is a master at keeping his head in the bucket. Because of his big floppy ears, they cover the whole bucket. The other calves don't have any open area to attack. I make him share with the new arrivals though.

If you have any farms that are split by roads, a lead steer is truly priceless to move them quickly across roads. I have a busy blacktop crossing that I have to cross once a month with 150 stockers. Ears makes it a snap to carry out though. I just call to him and the whole herd follows him up to the gate. I make sure the whole herd is accumulated behind the gate then open the gate and walk across the blacktop into the other pasture. The stockers look like bullets shooting across the blacktop. It takes about 30 seconds to move 150 head.

Ears is also a major tool on loading day. He will bring the whole herd into the corral or wherever I want to take them.

When you graze hundreds of calves, some will get sick. Some will even die. I absolutely hate to lose a calf. It is a sickening feeling. Even though I don't own them, it still bothers me terribly to lose a calf. You have to accept it, learn all you can from it, and go on. The first two weeks after you receive a group of calves, watch them like a hawk. You have to learn how to detect sick calves early and treat them immediately.

When you walk out to check the newly arrived calves, look for any calves that may be hanging back from the main herd. A lone calf not eating and standing by itself is almost certainly sick. Remember they are herd animals. If they are standing alone, that is not natural.

Other signs to look for are calves that move without purpose when you walk up to them. Their ears will be laid down. Their eyes may have a slight glazed-over look to them.

If a calf is lying down and looks sick, walk up slowly to it and make it get up. If the calf stretches when it gets up, it probably is not sick. If it does not stretch, but stands there, or walks off with its head down, or the back may even have a slight arch to it, this calf is sick. If you see a calf that you suspect may be sick, watch its head for several minutes from a distance. Its head will gradually lower down, simply because they don't feel like holding their head up for very long.

> *Learn to detect sick calves early and treat them immediately.*

Rainy cold periods will sometimes bring on sickness in

calves. Their hides get wet, they're cold and this brings on perfect conditions for calves to contract pneumonia.

Mud is one of your worst enemies. Do everything possible to keep them out of it. If a calf gets mud imbedded in his hair, it is impossible for him to keep warm. Health problems will follow. I move them often in rainy weather to keep them out of the mud. It also gives me a chance to really give them a good looking over while moving them. Keep an extra close eye on them during long periods of cold rainy weather. The earlier you can detect and treat a sick calf, the quicker that calf will get well. The earlier you can treat a sick calf, the better that calf will perform for you in weight gained.

Do everything possible to keep calves out of mud.

If a sick calf is not detected and treated early, then the medicine will not be nearly as effective. A human is the same way. Once you're sick and don't seek immediate treatment, it may take a month or more to completely rid your body of the bug. Once the calves' lungs are damaged by a virus infection, they will never gain weight like they should.

Pink eye can be a problem with grazing stockers also. I have found that with intensive grazing the fly problems along with pink eye are severely reduced. The reason is, with a fast grazing rotation you are leaving all the manure piles and fly eggs behind you.

Another valuable asset in preventing pink eye is providing stockers with quality water. Don't let them stand, urinate, and manure in water that they are drinking from. Remember, the younger the class of livestock you're grazing, the better the quality of water you need to provide for them. An old cow can drink some awful brackish water and still perform, but you're asking for trouble if you make a young group of stockers drink putrid water.

Most of the calves that I graze are weaned and have received both sets of shots. This really keeps the incidence of sick calves at a minimum. But I still have some calves that get sick. I strongly suspect most of the calves that get sick are a result of handling, shipping, and social stress they have encountered before they reach my farms. I am constantly working to improve my receiving and handling methods.

You have to teach yourself to go slow and not be in a hurry when you're around your calves. I'm not a Bud Williams yet, but with constant practice one day I may be close.

Watch your emotions around your calves. I know this sounds silly, but if you're in a bad mood, your calves can detect it and you will place stress on them. Watch a dog when you scold it for doing something wrong. It will tuck its tail between its legs and cower down or run off. Cattle are the same. I know it can be frustrating sometimes when working cattle and it is relatively easy to lose your cool. The stress you have placed on yourself, you have also placed on the cattle.

Remember they are animals of herd instinct. Once they become separated they will become very unsettled and stressed.

The handling facility for doctoring calves should be functional, but not elaborate. For the actual holding and sorting area, sometimes I use four to five strands of hi-tensile wire as the corral perimeter. This is the cheapest corral perimeter fence that I know of. I have never had a calf or cow go through it. Fastened to the hi-tensile corral, I will have several pipe gates that form an alley and a head catch at the end mounted on two posts.

> *The younger the class of livestock, the better the quality of water you need to provide.*

I form a V shaped funnel leading into the corral for penning the cattle. The V shaped funnel is simply one strand of hi-tensile wire nose high that is 100 feet long, which leads up to the entrance of the corral.

A lane that is hooked from your corral to all of your paddocks is a huge aid in separating out a sick calf from the herd. Once they are walked into the lane, it is a direct path to the corral. Bring several calves with the sick one if possible. The calves will handle better as a group, which means less stress will be put on the sick calf also.

The handling facility for doctoring calves should be functional, not elaborate.

For doctoring calves I also invested in a Cap-Chur™ air rifle that shoots an aluminum dart. I can administer up to 15 ccs with a single dart shot. For farms that don't have handling facilities, the dart rifle is priceless. I doctor the calves by shooting them from the pickup window. Most of the shots are 40 to 50 feet away. The rifle is powered by two CO_2 cartridges that propel the dart. When the dart hits the calf, the impact causes a powder charge to go off inside the dart which pushes the medicine into the calf.

Always shoot the calves in the neck. This is where the cheaper cuts of meat come from. Never shoot a calf in the rump where the prime cuts of beef are located.

One advantage of using a dart rifle is actually doctoring the calf on pasture without him even knowing that he has been doctored. The calf has no stress put on him from sorting and being placed in a head catch. A lot of the time the calves just walk off after getting shot. Sometimes they will run for a short distance as if a bee stung them. None of the other calves are stressed because there is no sorting.

When you are getting paid by the pounds of gain that you can put on an animal, the more docile and quiet you are around your calves adds up to more money in your pocket.

Animals that are worked through a sorting and handling facility are having pounds of gain stressed off them. You may only be doctoring one, but the whole herd may feel the stress of being handled, depending on your stockmanship and handling facilities.

Some tips for using a dart rifle are practice, practice, practice. Set up a hay bale and fasten an old rug to it. Tie a paper plate with a one inch black bull's eye in the middle of the bale. Step off several yardage markers. You can fill the practice dart with water, which represents the weight of the medicine. You want the dart to drop about a foot in trajectory. This means that you aim a foot above your target. The dart needs this arch effect so that the dart does not bounce out on impact of the animal. If the dart bounces out on impact of the animal, you are too close.

I will never forget the first heifer I tried to doctor with my rifle. I practiced about 20 minutes and thought I was ready to doctor. Well my first shot hit the heifer right square in the center of her plastic ear tag, a very humbling experience. That heifer looked like it had an earring hanging from its ear.

Chapter 19
MiG Promotes Wildlife

It is a true wonder to watch what happens with wildlife on idle land when good grazing practices are put in place. What I am talking about of course is Management-intensive Grazing.

I love to see a covey of quail flush in front of me and sail into a brushy protected area. I usually see deer or turkey when I'm moving stockers in the morning or evening. It is a very satisfying feeling to know that your management is causing this huge increase in wildlife. I personally think that the more paddocks you have, the better the possibilities of having an excellent wildlife habitat. It also benefits the wildlife the more tender and palatable that you can keep your forage. I've noticed where I have the best forage is also where most of the deer and turkey are seen eating.

All of my landowners like to see wildlife on their farm when they walk over it. I concentrate on giving the wildlife what they need to proliferate. Any time I make a grazing decision, I ask myself what effect will it have on the wildlife?

Just by the placement of hi-tensile wire you can make some very nice wildlife cover. I will build brush piles for rabbits when clearing brush. I place several large logs on the bottom, then crisscross two more across the bottom ones. These are covered with an old car hood or a piece of tin. The metal is covered with a layer of brush. The metal provides a good dry place for baby rabbits to dry off in the cool wet spring weather. A baby rabbit will die of pneumonia fairly easily if they don't have a place to dry off. The metal also keeps the base material from rotting so quickly.

The quail also like the dry area of the brushpile. The big logs on the base serve several purposes. They keep the brushpile from collapsing on the ground and also give the rabbits a place to get in that coyotes and dogs can not.

Quail need clumps of grass to nest in and for protection from predators. So when you are fencing off paddocks leave some scattered protected grassy areas around the edges of fields, ponds, woods, and old buildings. It takes roughly 40 acres for a covey of quail to exist. The quail seem to do well with MiG if you can give them a little elbow room. I love to hear a bobwhite whistling off in the distance while walking to move a herd of calves.

All timber is fenced off to prevent cattle from grazing in the woods. The timber is a very valuable resource to ruin with cattle tromping and compacting the forest floor. Once the forest floor has had all the leaves eaten off, then tromped and compacted, the rains just wash huge ravines and expose the tree roots. This all slows down the growth of the valuable marketable timber. This leaves nothing for the wildlife to eat.

> *Keeping cattle out of the woods keeps manure piles on the pasture where it belongs.*

Deer are browsers. They must have twigs, sprouts, acorns, etc. It takes 40 acres of timber to equal one acre of grass in grazing capacity. This one fact alone should tell you that it is not economical to let the cattle destroy your woods.

If cattle are allowed access to timber in the fall, they will clean up every acorn in the woods. The acorns are very high in tannin. They are toxic to young calves. The calves' rumen can not digest large quantities of acorns. You can tell if a calf has been eating large quantities of acorns fairly quickly. Their manure will be whitish and they may even appear to have diarrhea. They will literally blast a white stream out when they try to manure. If you see a calf doing this you have troubles. You have to get the calf off the acorns first. Then try to get some bugs working in his rumen again.

Managing leased land for wildlife in unison with your grazing operation pays back both emotionally and economically.

There are some bolus tablets that you can give them that may help solid up the stool somewhat. Talk to your vet and get his advice about what kind of medicine to treat him with. The best medicine I know of is to keep them out of the woods in the first place. The calves will do better, you will make more money, and have a cheaper vet bill.

I know cattlemen who lose calves every fall from acorns, but they refuse to fence the cattle out of the timber. One dead calf would pay for a lot of hi-tensile wire. Cows can eat a bucket full of acorns and get by, but not stockers. The deer need the acorns to carry them through the harsh winter weather in my area a lot worse than the cattle do.

One of the biggest economical advantages of keeping the cattle out of the woods is that you are keeping the manure piles on your pasture where they belong. That manure pile deposited down in the trees is not going to grow one sprig of grass. I look at every manure pile as a dollar bill. It upsets me to see all those piles of dollars lying under the trees where I can not use them to grow anything.

The cattle are transferring all your hard earned nutrients off the pasture and then you have to reach into your pocket for cash to replace these lost nutrients. This is a no win situation for the grazier, the land, the cattle, the wildlife and the trees.

I run one strand of hi-tensile around all wooded areas. I leave enough room on the timber side of the wire to run a brush hog every few years. This promotes sprouts, which the deer love. I can also cut the nuisance trees away from the marketable cash trees along the edge of the woods. This is called edge habitat. Wildlife absolutely thrive in edge habitat. 70 to 80 percent of their food is found in the edge habitat. So concentrate on providing all the quality edge habitat that you can on each farm.

I have one farm that is owned primarily for deer hunting. Before I leased the farm for grazing, the landowner told me they were lucky if they saw one small buck during the whole deer season. They saw very few deer, period. The farm primarily consisted of overgrown thatch of dead fescue and broomsedge. This is not exactly prime wildlife food.

> *People are starved for the opportunity to be associated with something that is natural and good.*

I overseeded all the pastures with red clover after the thatch was grazed off. Now all the pastures have 30 to 50 percent legume content. All the timber is fenced off, preserving the acorn crop for the deer and turkey. The woods are also starting to fill back in with sprouts, which give the deer quality browse and secure bedding areas. One of the problems with this farm was lack of thick protected cover for deer bedding areas. This absence of bedding areas caused most of the deer to find cover elsewhere.

This farm has 28 permanent paddocks that I rotate the stockers through. The diverse forage species include timothy, orchard grass, lespedeza, switch grass, redtop, red clover, alfalfa, hop clover, sweet clover, fescue, and blue grass. The deer are absolutely in heaven from having this constant tender salad bar to browse on. They actually get to take the very best cream of the forbs and sprouts crop that are present on the entire farm. This constant high protein diet is starting to grow some huge bucks.

A hunting lease may be possible on some of your leased land.

This past deer season the landowner saw seventeen deer opening day of deer season. The landowner harvested a nice trophy nine point buck. He saw four other nice bucks besides the one he harvested. I have mapped out a strategic area of the farm that is a perfect spot for a wildlife plot. The landowner is supplying all the materials and I am performing the labor to put in two acres of whitetail clover. I will lime, fertilize in the winter, and frost seed this spring. The patch will be flash grazed every month through the summer to keep it tender for the deer. I get the benefit of being able to mow the clover patch every month with my stockers and the landowner gets the chance of harvesting a nice trophy buck this next fall.

Your ability to manage leased land for benefiting wildlife and allowing it to thrive in unison with your grazing operation will pay you back both emotionally and economically. Your pay check emotionally is seeing the positive results of your grazing management on wildlife. I always get a warm feeling when I see wildlife enjoying the land that I have made into a wildlife paradise. It is a great feeling and one of my biggest bonuses for leasing idle lifeless land. I highly recommend it.

Economical payback from managing wildlife in concert with grazing livestock usually occurs in the form of long cheap leases on the land. People are starved for the opportunity to be associated with something that is natural and good. Many landowners value seeing wildlife on their land more than they value a rent check. I urge you to learn as much as possible about what it takes to hold different species of wildlife on land. Check with your local wildlife conservation department for information on specific information on each desired wildlife species in your region. They should have some brochures that give a full breakdown on what it takes to draw and hold each species on your leased land.

One of my magical moments with one leased farm happened one night while moving stockers. I was on my way to the back of the farm to move calves when I saw what looked like a huge house cat with an even bigger extra long black fluffy tail. The animal seemed to just flow across the top of the ground. I immediately knelt to the ground and waited as the animal came closer to me. It was completely unaware of my presence as it made its way down the trail towards me. When it rounded the curve of the trail, it was headed right for me.

I was shocked with amazement. It was a beautiful gray fox that had a small rabbit in its mouth. His huge fluffy black tail looked as long as his whole body. He stopped twenty feet directly in front of me and just stared at me. I never moved a muscle and neither did the fox. We both just stared at each other for a while. I had never seen a gray fox in the wild, not to mention twenty feet in front of me. The gray fox knew this

Prime hunting areas are vanishing quickly.

clump (me) didn't belong on the trail. It finally wandered back the same direction from which it had come.

Since then I have seen gray foxes on other occasions on this same farm. One night I heard an eerie scream. It sounded like a woman screaming bloody murder. It brought the hair up on my neck to hear such a wild sound coming from the exact spot where I always saw the foxes. I had always heard that a gray fox screaming sounded nightmarish. I'm a firm believer now. The reasons the fox are able to thrive on this farm is because of the brushpiles that harbor tons of rabbits, voles, and mice.

You don't want hunters on the land every weekend.

A hunting lease may also be possible on some of your leased land. It is getting increasingly popular around the country to lease land for hunting purposes. Hunters like the security and freedom of hunting on land that has no other hunting pressure. It is a nice feeling to go out for a leisurely hunt and not have the worry of someone else hunting in the same spot where you are. First of all, it is dangerous to hunt within bullet range of another hunter. It is also a sick feeling to be sitting in a prime deer spot and when it gets light outside all you can see is orange vests around you. That's not exactly a safe situation to be in.

Prime hunting areas are vanishing quickly. They are indeed a rare valuable commodity that you may be able to utilize. Land may have a pond or two to offer. Make it a hunting and fishing lease.

The more you are willing to provide to the hunters the more you can charge. You may have an old home site or cabin that the hunters can stay in. Water and electricity raise the hunting lease even more. The more comforts the better. Once you get a good set of ethical hunters who pay on schedule and appreciate your land, take care of them.

I have had several sets of troublesome hunters in the past, which makes me appreciate the wonderful ones that I have now. I will make an attempt at giving you some warning signs to look for when screening hunters for a hunting lease.

First, to get a list of hunters to pick from put an ad in a paper that serves a large city that is at least 100 miles from your land. The reason for the large city is that it gives you lots of exposure to city folks who have no chance of hunting on private land. They usually have a larger income than most country folks do.

The reason you want the city that you are going to advertise in 100 miles from you is that it keeps the hunters from being on the land every weekend. If they have to drive a 200 mile round trip to hunt or fish they are not going to do it every weekend. They will appreciate the outdoor experience more each time they come out, simply because it cost them some pain to get there.

I appreciated every second of my elk hunt several years back after driving 2000 miles to get there. It is the same with a simple hunting lease if they have to drive a good distance to get there.

Set up an appointment with the prospective hunter or hunters. Never allow more than three hunters on a lease. Never allow them to bring anybody with them, not even their wives or girlfriends. Inform them that you want to walk over your land with them to see if it is the kind of hunting land that they were looking for. If they act like they are in a hurry and do not seem eager to walk every acre of the prospective land, this is a warning flag. If they are not willing to spend some time and walk the land, or talk with you about themselves, then they will not appreciate your land or your wildlife.

Look for hunters who love wildlife and harvest it responsibly.

You can tell a lot about a person by simply walking and talking for a couple of hours. If the hunter excessively brags about his previous hunting endeavors, pass on him. Watch his reaction when you jump a deer while walking the farm. Does he marvel at it or does he exclaim how easy he could have killed it?

I look for hunters who love wildlife and are responsible about how they harvest it. You can tell a lot about what kind of hunter a guy is by listening to him tell a previous hunting experience.

I signed up three hunters one fall for deer season. I think I made every mistake that a person could in the process of picking them. First of all, I waited until three weeks before deer season to place my ad. You do not want to be hurried when selecting the right group of ethical hunters. I must have had 50 calls with my first ad. I was so overwhelmed with calls that I got tired of answering the phone.

I picked these three young guys in haste and paid for it. After their first walk, one of them was in an extreme hurry to get back to town after briefly looking at the hunting land. This same hunter also shot a red squirrel off the tree limb in front of him on opening day of deer season. He bragged about it after he got back to camp, how he blew it to pieces. These three young guys were all supposedly lifetime hunting buddies. I noticed that they were very proud of their previous hunting experiences and seemed to enjoy picking on each other a little bit. When they showed up for opening day of deer season, one brought his wife and the other two brought their girlfriends. All the women got in a fight and refused to stay the night unless the guys made up separate sleeping arrangements for each of them.

A hunting lease can provide a nice source of income.

The next day the guys got drunk and got in a fight. They were the hunters from hell. Needless to say, that deer season could not end quickly enough for me. It was a great learning experience for me to go through though. It was a terrible thorn to sit on, but I got through it. The bottom line is do not get in a hurry when screening potential hunters for your farms. Remember, the harder the lesson, the deeper it brands it into your soul.

The hunters I have now were selected in the off season. I took my time getting acquainted with them and explaining the exact terms of the hunting lease. I substantially raised my price for the hunting lease and also got a signed six year hunting lease. By raising the hunting lease it helps screen out the bad ones. They treat my farm and wildlife with respect. They are friendly, honest, hard working people who like to hunt and fish.

They pay their hunting lease every January without being asked. They call me when they are planning to come up for the weekend. I have helped them put out some food plots that are placed at strategic areas around the farm. They enjoy harvesting the venison and eating it as well. I provide them a safe prime hunting area to do exactly that and have developed a strong relationship in the process. A hunting lease is a nice source of income that may be practical for some of your leased farms. Check it out.

Chapter 20

Timber Stand Improvement

Timber stand improvement (TSI) is a service that you can offer potential landowners for their wooded areas. Most landowners care about their trees, so if you come along and offer your assistance in improving what is already there, so much the better. You're probably wondering what does timber stand management have to do with being a grass farmer?

Well first of all, it is a very good selling point for securing an economical lease on a prospective idle tract of land. It does not require a lot of investment in tools to perform TSI. It does not require an enormous quantity of time either. It is all tied together folks. If you show the landowner that you are concerned about their timber, this may be the very thing that sways them in their decision to lease their ground to you instead of the next guy.

TSI is a very economical reward for the landowner as well, in the form of potential timber sales. Most landowners do not know how to manage their timber tracts. If you treat timber like a crop, it is a very economical resource. By managing the timber areas correctly and harvesting the timber like it should be, the landowner can get a timber sale every seven to ten years. The nice part is each progressive timber sale is more economical than the previous timber sale as long as the timber harvest is managed. I cannot over emphasize the importance of a well managed timber harvest.

The old belief was that a timber owner might get one good timber harvest in his lifetime. This is not the case at all if you learn to perform timber stand improvement practices on the wooded areas and supervise the timber harvest. Basically what TSI involves is protecting the trees from grazing and removing the invaluable trees so that the marketable trees can express themselves without any competition for sunlight, moisture, space and nutrients.

For the trees to grow the fastest, the crown (top) needs full exposure to the sun. The tree top covered with leaves is a huge solar collector, which is very much the same as a blade of grass. This is where you come in. By walking through the woods and removing the nuisance trees, you are releasing some very valuable stems (trees). In my area, oak and walnut are the money trees. These are the trees that I concentrate on growing to their fullest potential. I cut all trees that are in direct competition with the money trees.

Carry some stump killer and a paint brush with you. Each potential money tree should have a 25-foot-free circle around it devoid of other trees. This guarantees the crown of the tree full exposure to the sun. When you cut a tree, paint the bark ring on top of the stump within three hours. You are wasting your brush killer if you paint the center of the stump, the bark ring is what feeds the tree. By waiting longer

> *If you treat timber like a crop, it can be an economical resource.*

than three hours to paint the bark ring, the sap in the trunk has drawn down into the roots and sealed over. By painting the bark ring immediately you will get a good kill on the stump. You don't want any competition coming off the stump and competing with your money tree.

One of the biggest growth retardants on walnut trees are grape vines. They spiral their way up into the crown of the tree and strangle the crown. I never realized this until a consulting forester explained and pointed it out to me. Next time you walk up to a walnut tree and it looks spindly and stunted, look for grape vines that may be strangling it.

I cut all grape vines on walnut trees and paint them immediately with brush killer. I also cut all limbs off the walnut trees with a hand saw that can be reached with a 16 foot extension ladder. The tree heals over where the limbs once were and makes a very nice extra long saw log.

Be extremely careful when performing this exercise. Have someone on the ground holding the bottom of the ladder. Always tie the top of the ladder to the tree and position the ladder perpendicular to the limb that you are cutting off. By cutting off the bottom limbs you increase the value of the potential log significantly. If you have two straight trees growing off of one stump and they are larger than 10 inches leave them both. If they are smaller than 10 inches, save the tree that has the most open and well developed top crown and remove its buddy. Trees will grow 30 to 50 % faster with these basic management practices.

When judging marketable trees, measure the tree's circumference at waist or chest high.

When it comes time to pick the trees to harvest, if you do not feel comfortable with your ability to pick out the proper trees, consult your local conservation commission. They have forest managers who will mark the trees to be cut for you free of charge. Get cash bids from four to five different loggers on your marked trees. Have a signed contract stating the terms of the log harvest.

Listed below are the essential topics that need to be included in the timber harvest contract:

1. Harvest only marked trees.
2. Log skidders can only work when the ground is solid.
3. The log loading area is to be clean when finished.
4. Half of the money to be paid at the time the contract is signed.
5. The remaining half of the money to be paid before any timber is cut.
6. Logger must have liability insurance on himself and his employees.
7. Logger agrees to limit as much as possible damage to unmarked trees.

Before you agree to any logger's bid, ask them for some prior customer references. If they refuse to give you any names, that is probably a pretty good sign that you do not want them to perform the harvest. Remember the old saying, "The best advertisement is a satisfied customer." If he is a reputable logger, he will not mind giving you some prior references. Just because a logger gives you the highest bid on the timber does not necessarily mean he is the best choice.

We have all heard people exclaim about how big their trees are and how much they ought to be worth. You do not want to harvest them too young, but the opposite is true also. Once a tree reaches a certain age it will start decaying. This process usually starts on the interior of the tree where it is not visible. The tree may appear to be perfectly solid from the outside, but may be actually hollow. This is the risk the logger takes when he bids on the marked trees.

One tip I learned from a consulting forester was that when judging marketable trees, never measure the stump portion of the tree. Measure the tree's circumference at waist or chest high. You can also hold a yardstick across the tree to get the width of the tree. This measurement separates the men from the boys when it comes to dollars. In my area a tree that measures 20 inches across at chest level is a very valuable saw log. A tree will always measure larger on the stump, which gives you a false reading as to the actual board feet in the tree, which is taken from the measurement at chest level.

Chapter 21

Using Cost/Share Programs

In some areas of the country there are programs through the state or federal government that offer cost/share assistance to implement programs that are environmentally friendly. These programs may range from establishing a permanent stand of grass on highly erodible land, fencing to exclude livestock from timber, pond construction, running water lines, water tanks, timber stand improvement, fencing to exclude cattle from streams, or constructing water tanks fed by springs. These are the major ones that are very beneficial for grass farmers.

One requirement on most of these practices is you must install some form of rotational grazing. There are different views from people concerning cost/share practices. Everybody has a right to their view. That is what makes America a wonderful place to live. I pay my fair share of taxes on everything I buy and earn. So I do not have a problem with utilizing cost/share programs if they are offered.

First, these cost/share programs do not pay the whole amount, but are strictly an incentive to get people to implement environmentally stimulating practices on their land. The programs benefit the ground and surface water quality, and hold the soil in place so that our children will have something better to manage than what we had.

I explain cost/share practices to landowners when these programs would be beneficial to their land. An example might be a farm with no water. By installing paddocks, the farm may qualify for a pond if it meets all the other requirements. Now all of a sudden that farm that looked impractical for livestock production because of the lack of water has all kinds of promise. The same results can be obtained by bringing up the soil fertility and ph level on a depleted farm. If the landowners express interest in implementing a cost/share practice on their land, then I offer to do all the paper work and physical labor to install the practice.

I had one landowner who wanted to do a permanent grass stand establishment program. This farm was so poor that it would only grow broomsedge. I did all the paper work and the farm was approved for the practice. The Natural Resource Conservation Service paid 70% of all the lime, fertilizer, seed, and drilling cost. The landowner paid the remaining 30% of the establishment cost. This gives the little guy a chance at getting a poor farm up and producing without bankrupting your savings account. This particular landowner is now so happy with his improved farm that he is now liming and fertilizing 25 % of the farm on a set schedule as soil samples dictate. He is a lot more in tune with his farm and sees the benefits of conserving and protecting his natural resources.

> *Cost/share programs are an incentive for people to implement environmentally stimulating practices on their land.*

Most landowners want to improve their property. This is an economical way for them to do so. This is another service that you can offer a potential landowner. Explain in detail to the landowner all the steps that will take place before starting the cost/share. In the cost/share practices that I have done, I explain to the landowner that he is responsible for the materials cost and I provide all the labor and fill out all the necessary forms. I explain to the landowner that he will be reimbursed the amount that is approved on the cost/share usually four to six weeks after all the forms have been signed and tickets calculated.

There is a lot of work involved in implementing a cost/share practice, but the rewards are worth it. The paperwork has to be done exactly like the instructions say. Keep good records and a file with all your receipts. Keep copies of all materials tickets for your own records. These are great reminders of exactly when the materials were put on the farm. Any items that you buy for the cost/share practice, have a signed and dated receipt showing your check number on the bottom of each ticket.

Make sure you do not start on the cost/share program until you receive a signed confirmation that you have been approved for the program. Make sure you understand the exact amount of each material to be used on the program. If the landowner wants to implement a cost/share program, just plan on six months minimum for getting all the paper work done first. If cost/share is something that interests you or your landowners, check out your local NRCS office or state conservation commission and see what programs are available.

Chapter 22
Managing Your Time Effectively

There is only so much time to get things done in a single day. You need to constantly monitor your daily actions to weed out the unproductive periods. Keep a record for one week on everything you do throughout the week. Record each activity every thirty minutes. At the end of the week total up all the time and place them in their respective columns. It may amaze you how much time of the week is wasted in unproductive activities.

There are only 24 hours in a day, we cannot magically produce more hours. So we need to learn to effectively utilize every minute that we have. The old saying, "Early to bed, early to rise," is a great place to start. If you sleep in every morning because you stayed up late watching television then you place yourself in a losing pattern of wasting the most productive part of the day. The first six hours of the morning are the most productive portion of each day. You wake up and are refreshed from a good night's sleep. The morning air is cool and your energy level is at the peak.

If you are working on a project and feel burned out, switch projects and do something different for a while. When you finish a project, reward yourself. Do something special that you enjoy. This is very important, not only as a treat for a job done, but a milestone that was reached after much work and sweat.

I see a lot of young folks who do not have a clue about managing time. When you are young time it is not that important. After all, you have your whole life ahead of you. But as you age, you have a greater appreciation for time simply because you do not have as much left to waste. When you wake up in the morning, be thankful for that day.

I hear so many people who make the comment, "I wish I was retired so that I would not have to go to work." These people are wishing their life away and do not even realize it. A lot of the people who are retired are bored to death with nothing to do. They lose track of what day of the week it even is. So do not think that retirement is the golden day that you are looking for. Right now, today is the day that you should be focused on. Make the absolute most of every day that you live. Make the chips fly, enjoy the journey and challenges of reaching your goals.

If you have a habit of turning on the TV when you come in the house, this is a habit that you should work on breaking. A lot of people don't even think about it. It's just a natural thing to do when you enter the house. Next time you come in the house and you walk toward the TV for the on button, have a really good book on grazing, soils, direct marketing, etc. sitting on top of the TV. Pick up the book, leave the TV in the off position and read. I

> *Do not waste time and energy trying to convince someone who is unwilling to learn.*

am convinced that you can never read too much. This is how you teach yourself.

I get a little aggravated at people who say to me, "Well you know all about that MiG stuff, grasses, soils, etc, that is why you have such good luck with your farms." Nobody force fed me all that information. I had a hunger for it and started reading every publication that I could get my hands on to learn everything I could on the subject of grass farming. The *Stockman Grass Farmer* is the single best publication of them all. If you're interested in improving your knowledge in the world of grass farming there is no other publication that covers it like this one does every month. You're missing the boat by not subscribing to it. I can hardly wait for each copy that I receive every month.

> ***Don't look for support from your neighbors when you start doing things differently from them.***

People waste a lot of time second guessing their decisions in life. All you're doing is backing up and wasting valuable time. Write down your goal and do not take your eye off it. Stay 100% focused on it and you will reach it. Remember your mind is your own personal king. Give it its rightful honor. Your mind is perfectly capable of doing anything you truly desire. You just need to keep the clutter out of it so that it can perform at its peak.

You know how your work-shop can get in several weeks if you do not put your tools back in their proper place after using them. Well, your mind works the same way. When it has a lot of clutter in it, it struggles to perform at its peak. Complete one task entirely. Strike it from your list and go on to the next. There is something to be said for the feeling of a job well done. If you're going to go through the motions of performing any task, do the very best job that you can. You're only cheating yourself if you give it a half effort. It really does not take much more time to do the job right the first time. Then you will not have to come back and do the same job again in several years. This really pertains to everything you do in life, not just grass farming.

Be aware of your surroundings. What I mean by this is don't get in a rut that everything around you is the way it should necessarily be. Examples might include some of the following:

1. Moving a watering site to a more bang-for-the-buck location, an area that will water multiple paddocks from one tank.
2. The same thing for paddock divisions – are the paddock divisions where they really need to be?
3. Grazing multiple livestock species, which may include sheep, goats, or pastured poultry to mix with the cattle. The more diversified you are, the better insurance you have against a sagging market. When one market is down another species market will be up.
4. Idle land that you drive by everyday around you that may be prime for leasing.
5. Owning heavy metal that is constantly keeping you hot trotting to the parts store while it is rusting away and keeping you in the poor house.
6. Are you making hay instead of figuring out how to let the cows harvest their own forage this winter right on the stem?
7. Fencing off timber so that you will have a cash timber sale instead of poisoned calves every fall from eating acorns.

Keep a list of possible action items that bother you about the way your operation is being run. Knock them out one at a time and remove them from your list. This way they don't seem so insurmountable. Get in a habit of always asking yourself if there is a better way to perform this function than is being done at the present? You will be surprised at the new ways that you start to look at things.

It is exciting to be different from the mainstream. You are a pioneer, doing things that are not accepted as being right. So learn to separate yourself from what is popular with the rest of the mainstream farmers. This is where your greatest opportunities lie. You have found an unfair advantage because everybody else thinks you're nuts for trying it.

Do not waste time and energy trying to convince someone who is unwilling to learn. If I explain logically what I am doing and they look at me like I'm nuts, that's fine. You will more than likely make someone mad before you change their mind, if their heart is not in the right place. What I mean by this basically is, if they want to change they will. If they are close-minded to new ideas they won't.

Don't go looking for support when you start doing things around your farm differently than your neighbors. Most of them are praying that you will fail so they can be the first ones to tell you, "I told you it would not work."

Learn to love the feel of the wind in your face, because you are charting uncharted waters. It's a wonderful feeling once you're out there meeting your challenges head on.

Chapter 23
Have Fun, Enjoy the Journey

Developing a successful grass farming enterprise is a wonderful journey. To be truly successful at anything in life you need to breathe, eat, drink, smell, swim, crawl, dig and be willing to do whatever it takes to make it happen. There will be times when you may be overwhelmed with doubt, fear, anxiety, just down right scared if you can pull off a tough goal that you are in the midst of tackling.

Don't sweat it. This is perfectly normal behavior when you're treading on uncharted territory. Anytime you do something that is off the beaten path, you will find events and circumstances that make you feel uncomfortable. How you react to these situations is critical and will determine if you succeed or fail in your endeavors.

I made a lot of mistakes when I first started leasing farms and custom grazing. I had never done either before. I still make mistakes. That is how you grow and prosper. By not being afraid of making mistakes and having a tough enough hide to weather the storms, I just keep going.

I had people tell me to my face that I was crazy, but this was like throwing gas on a fire. It just made me even more determined to reach the goals that I had set for myself, regardless of what other people thought. So learn to accept criticism with open arms. It truly is a good thing if you are doing something new that makes other people uncomfortable. The way most people react to something new is they attack the idea and the person who actually has the gall to try it. There is nothing that makes them madder than for them to say it cannot be done and you go out and do it!

Mistakes are how you grow and prosper.

There are people whom I have met who are very uncomfortable talking to me about grass farming. They are so entrenched in the old mindset of ranching that to even mention that there may be a new more economical way of ranching simply infuriates them. They will seek out other people who have the same mindset as they do. This makes them feel comfortable about what they are doing. They will sit around in their circles and talk about all the hay they put up, their new feed bunks, new tractors, balers, you name it. As long as they can find someone else who is doing precisely what they are doing then they feel good about their management skills.

Remember, there is safety in numbers and this is where your weakest link may be. You will not have the luxury of being surrounded by people doing exactly what you are. If you can envision in your mind the actual event happening, it will happen. Your mind is your most valuable asset. Never under estimate the power of your precious mind. Anything that you desire or truly enjoy ensures that you will be successful at it if you give it everything you've got.

To be truly successful at anything in life, first of all you need to love doing it. You will work harder and be more determined to reach your goal if it is something you love dear to your heart. It's not near as dreadful putting in a 12 to 14 hour day working if you really love what you're doing.

I can build fence all day putting in paddocks and just be happy as a kid in a candy store. I know that at the end of the day there will be some impressive looking paddocks to rotate calves through. Anything that I can do to take full advantage of every beneficial grass plant just flat turns me on.

You're probably thinking this guy is nuts. Well I will admit I am a grass grazing nut. It just plain gets me excited turning previous idle pastures into profitable low cost grazing systems.

Don't expect to have a cake walk when you start out leasing land. There will be some tough times that will challenge you as to how determined you are at reaching your goals. These are simply bumps in the road that you drive over and keep going toward your destination.

Every problem that I have faced so far I simply do not allow it to be a show stopper. I note the mistake and go on down my path toward my goal. There is nothing in this world that can stop you from reaching your goals if you keep on trying. There is always a solution for every problem you encounter.

I make a list of the possible solutions for each problem that I encounter and study them for a while before picking one. By writing them down on paper they are easy to focus on independently of each other. I will pick out the most economical approach first, the quickest to implement comes second.

> *There is always a solution for every problem you encounter.*

I had a classic problem arise on one farm that I leased several years back. I had secured a lease on an abandoned farm that basically had no fences to speak of. The farm had great potential though – lots of grass, several water sites, a cheap lease, and great landowners. The problem started when I accepted a herd of cows to graze at a set turnout date before I had the farm completely fenced. At the time that I accepted the cows, I had several months before they were due to be delivered. I figured it would be no problem to get the whole farm fenced with hi tensile in that amount of time. Well, the weather turned bad and fence building stopped. There was a lot more brush to clear than what I had originally calculated. The cattle owner was depending on me to take his cows on the date that we had agreed on and had made all the necessary contacts to do so. So I had to sit down with a pencil and write down a list of solutions. My list follows:

To be successful you have to love what you're doing.

1. Call the cattle owner and tell him that I would not be able to take his cows to graze at the agreed time.
(I did not like this solution. I had already given the cattle owner my word and had a signed grazing contract. Also this was not exactly the best way to get your name spread around as a reliable custom grazier.)
2. Hire some outside help to get all the fence erected in the short amount of time that I had.
(I had no extra money at this time, so this solution did not look very attractive to me either.)
3. Work like a dog and pray that I got all the fence up in time.
(This was not going to happen. There was simply too much fence to erect in the length of time that I had.)
4. Fence the largest grass area of the farm first that had a dependable water source.
(I liked this solution a lot. This would allow me to graze the cattle on the date that I had previously agreed to.)

I picked solution number four simply because it solved all my immediate problems. The solution was quick and made the most sense for the time schedule that I was facing. By focusing on the section of the farm that had the best forage and water, I reduced the immediate fencing job by 75 percent. I sat down and drew out where I wanted my fence line to run and got it fenced with time to spare.

Once the cattle were delivered, I continued to fence in additional sections of the farm to graze. By mid summer the entire farm was fenced into sixteen 10-acre paddocks. Once the cattle were on the farm, their grazing patterns gave me a good idea about where to run the new paddock divisions.

This example is pretty typical of the kind of problems that you may run into. But by systematically going through each solution to see which made the most sense, the problem was solved without any real discomforts or setbacks.

You want to be careful and not simply react when faced with a problem. Take some time and analyze your solutions before making any decision. I guarantee it will pay you enormous dividends and give you a huge sense of satisfaction along with a growing confidence to face your next road bump.

You will not become a grazing or leasing Einstein overnight, or for that matter even in several years. You may have to pay your dues in the form of bumps, bruises, scrapes, or what some people call the "The Learning Curve."

I remember back when I was a teenager, how I thought all the older people were just plain stupid. I knew everything there was to know about life and nobody could tell me any different. I look back now on that teenage stage of my life and I see who the really stupid one was. I'm still in the learning curve. I've developed more wisdom than I

> *Take time to analyze solutions before making decisions.*

had as a teenager but I still need to keep exercising my brain every day. Life is the learning curve, so enjoy it to it's fullest potential everyday that you breathe. You can never get back time, so don't put off what can be done today for tomorrow.

Life can compare to a chess game in lots of ways. Each player gets the same number of pieces at the start of the game. The object of chess is to capture the king from your opponent. So when you make a move on the chess board, you never want to move a piece backwards or be forced to move backwards. Anytime you move backwards, your opponent is gaining a stronger force against you. You wasted your move and now your opponent starts to exploit your weakened defense. Pretty soon your opponent has you running all over the board and you're just trying to keep from losing your scalp. Comparing these same situations to life gives you some perspective as to how easy it is to go backwards if you don't watch your moves closely.

Whatever you want in life, dig down and go for it.

For example, if you watch a couple hours of TV every night, what's that doing for your goals? Absolutely nothing but allowing your opponent to gain ground on you or capture some of your ground (your goals) that you had your eyes set on. Instead of watching TV you could be reading a knowledgeable grazing book, calculating your grazing plans, designing marketing plans, etc. Any time in life that you are not moving forward you're coasting in life, allowing your competition to gain an offensive position on you. I would rather play offense than defense any day.

You may be thinking, "Heck this guy is talking like life is full of offense and defense." I figure if you're going to play the game (full life) you might as well play it to win. What rewards are out there for people who give half an effort to everything they do? I really do not want to know what their rewards will be. I'm focused on the whole reward, which comes from giving everything I have in my body. You will never feel like you cheated yourself if you give 100 percent effort in everything you do in life. Wonderful things will start to happen to you.

I have had people tell me numerous times, "Greg you are the luckiest person I know." Guess what? Luck absolutely has nothing to do with it. You make your own what-other-people-call "Luck" in your life. The more you want something the more luck you have in your life. What a coincidence. Do you see a correlation here? I think I've made my point. Whatever you want in life just dig down and go get it. Enjoy the journey my friends.

Chapter 24

Leasing and Grazing Stimulates Local Economy

By leasing land and custom grazing cattle on what once was useless idle pasture ground you are also providing a huge boost to the local economy in many ways. First, the local sale barn has more calves to sell and buy as a direct result from your grazing activities. This brings in extra buyers who are looking for large uniform groups of calves that have been grazed together.

Something that I am starting to do for my cattle owners is attend the sale of the calves that I just finished grazing for them. Tell the auctioneer that when they run the calves in the sale ring you would like to give a brief grazing history of the calves.

The comment goes something like this, “These 100 calves grazed as one group all summer long. They were moved to fresh grass every two days and never were sick (if that was the case). These calves are all used to each other so you will be buying a group of calves that will have no social stress or pecking order to establish. These calves were handled every two days, so they are used to people and easy to handle. They are ready to eat and gain for you as a group."

The cattle owner will usually have a big ole grin of satisfaction on his face when you finish this speech. The cattle buyers are licking their chops to get their hands on this group of calves and will bid up and give a few dollars more per hundred weight on a group of calves when they know their history.

When you're talking about 100 seven-weight calves at two dollars more per hundred weight, that is some serious extra cash in your stocker owner's pocket. That comes out to an extra $14 per head, $1400 for the whole group, simply because you told about their positive grazing history. You're also filling up a pot load (Semi Trailer) very quickly with large groups of calves. Cattle that are grazed together as a group perform better in the feedlot as a group.

Be sure to mention this marketing sale service to the stocker owner when you are trying to secure a group of calves for grazing. This extra marketing service may be enough to sway him to give you a try in grazing a group of calves for him.

The local cattle haulers make some nice income as a result of my grazing business. They will come in with their large trucks and trailers to haul the calves to market and to the farms to graze. I buy all my mineral from local dealers for the entire year in the spring. The stocker owners reimburse me for all purchased mineral. The mineral dealers give me a nice discount because I purchase a large volume at one time. This helps the individual stocker owners hold their mineral costs down. I can buy 15 tons of mineral cheaper than they can buy a half ton of mineral for their own calves. This is a nice savings you can pass along to the stocker owner.

A little extra marketing at the sale barn can bring added dollars.

All my fencing supplies are purchased from local fencing dealers. Any medicine is purchased locally at farm supply stores. All feed supplements are purchased at local stores. All hay is purchased locally. All pond building services are done by a local excavation company. All legume seed is purchased locally. All lime and fertilizer is purchased locally.

All these companies are feeding off the successful grazing company. It makes for a very strong and diverse local economy. Everybody makes a little money from your original idea of building a grazing company from leased land. You may also choose to invest some of your profits in developing additional leased grazing land that also continues to produce more local commerce for the community.

It gives me cold chills to think what thousands of operations just like mine could do to the local rural economy nation wide. Heck, we wouldn't need to have all these government-supported $300,000-per-year farmers. The farmers and local merchants could support themselves by their own actions. I believe this is what everybody wants deep down inside anyway. Nobody wants a free handout. They would like to have a chance to earn their own way. It is in our own nature to be independent. It's very rewarding to receive a nice income from your own original idea, but also to see other people benefit from your actions as well.

An extra $14/head equals $1400 for a group of 100 calves, simply from explaining their grazing history.

When I drive around and see idle pasture land, I see a great opportunity not being taken full advantage of. I cannot put it in any simpler words than that.

When talking with prospective landowners, make sure

you cover the subject of how your grazing practices have a positive effect on the local economy. Not only is he improving his land, he is also involved in helping improve his local economy by being a part of this grazing practice in the form of him providing the idle land to graze. I think it is very important to let the landowner know that he is helping his local economy by doing something with his land besides letting it lay idle. Not everybody gets a chance to help other people in the community and by pointing this out they feel like they are helping out right where they live.

It makes me sad to drive through all the small towns that have basically dried up from not having any local commerce. All the stores are empty, no merchants anywhere to be seen. What used to be a thriving small community now looks like a ghost town. All the small farmers who used to patronize the local merchants are gone. A lot of the land is still there, but not being used.

Chapter 25

Keeping Accurate Grazing Records

To keep your grazing company running like a well oiled machine you should keep accurate grazing records. These records are invaluable when you need to refer back to them in the future.

I keep the records as simple as possible. This prevents confusion and makes them more user friendly. I keep separate grazing records for each farm that I graze. This allows me to keep an accurate record of the expenditures and income generated by each farm every year. I use an Excel spreadsheet to keep all grazing records simply because it was the first software I learned.

You're probably thinking, well that is great, but what if you don't have a computer?

You do not have to own a computer to keep good records. It just takes a little longer to record your data. Simply copy the charts that I have on the following pages or design a chart to fit your own needs. By all means use records that you are comfortable with.

If you are grazing calves from several different owners and are grazing them on different farms, it is very easy to get numbers confused. This is where a good chart that tracks each month of grazing is priceless. At the end of the month, it takes about 20 minutes total to fill in all the data for all farms. I send out a copy of the calf grazing days, which has the total bill for the month to each stocker owner.

On each farm a chart is started when a group of calves is turned out for the grazing season. At the top of the chart is the farm identification. (See page 220.) On this chart I record the date the calves were unloaded and the name of the driver who delivered them. I also record the number of head on each load, sex type, their average weight, their group weight and the total number that are on that particular farm.

You may have numerous loads delivered throughout the first month to get the farm properly stocked. By recording these individual loads of calves that are brought into the farm it also allows me to keep an exact count of total calf grazing days that each farm provides for the entire grazing season.

After the farm is stocked I have a column that has the total group weight at the time grazing begins. This is your grazing turnout weight that will be used to subtract from your end of the grazing contract load weight. The difference in pounds is what you get paid, so by all means it is very important not to get these numbers mixed up.

If a calf dies during the grazing season, the average spring turnout weight is used to subtract from the total group weight.

Along with the monthly grazing chart page, (just described) I keep a separate chart that keeps monthly and yearly totals for the farm. (See page 222.) This keeps all the grazing data from each farm on one sheet of paper.

Depending on the grazing contract with the stocker owner, sometimes it is agreed on the contract to pay a half pound of gain per calf grazing day for each month.

Farm Identification: Smith Home Farm July Grazing Record

Date Unloaded	Name Of Driver	Number Of Head	Sex Type	Average Weight	Group Weight	Total Number	Calf Grazing Days
03/25	Leon	77	heifers	577	44,405	77	
03/29	Steve	35	heifers	570	19,965	112	
04/08	Joe	18	heifers	552	9940	130	
04/16	Bill	6	heifers	610	3660	136	
04/21	John	26	heifers	534	13,886	162	
July death loss	1 heifer	subtracted	566 lbs			161	
Total Group	**Weight**	**As Of**	**07/31**	**90,680**	**pounds**		
Total Group	**Running**	**Average**	**Weight**	**Equals**	**566 lbs.**	**04/21**	
Total Grazing	**Days From**	**March to**	**October**	**Equals**			**31,715**

445 Total Head (ALL FARMS)
El Ranchero's TOTAL CALF GRAZING DAYS
July 1999

Farm Description	Total Calf Days For Month Of May 1999
Smith Rucker Farm (147 Steers- Spring 1999)	4557
Colley Home Farm (161 Heifers- Spring 1999)	4991
Lewis Farm (117 Heifers-Spring 1999)	3627
Titus Farm (20 Heifers- Spring 1999)	620
Grand Total	**13,795**

13,795 calf grazing days x .16 cents equals $ 2207.00

Grand Total Calf Grazing Days 1999

Date	Month	Year Total
Jan-99		
Feb-99		
Mar-99	532	**532**
Apr-99	4074	**4606**
May-99	5022	**9628**
Jun-99	4860	**14,488**
Jul-99	4991	**19,479**
Aug-99	4991	**24,470**
Sep-99	4830	**29,300**
Oct-99	2415	**31,715**
Nov-99		
Dec-99		
Year Total	**Total =**	**31,715**

I try to work with the stocker owner as much as possible on whatever he is comfortable with in the form of a payment schedule. Some stocker owners may not have the cash reserves to pay you a monthly fee, so I will put off my payday until the cattle are sold.

Remember, you have to be flexible to make custom stocker grazing work. I constantly try to put myself in my stocker owner's shoes when it comes to payment agreements. The stocker owner does not get paid until they sell their calves. I am not insulted at all if a stocker owner wants to postpone grazing payments until he has sold the calves. On most of my grazing contracts I am paid .16 cents per calf day. By paying some of the grazing bill on a monthly basis, the stocker owner is not hit with such a large bill at the time they are sold.

Getting back to the monthly grazing chart, it is very simple to calculate the monthly grazing bill for each set of calves. You simply multiply the days of the month by the number of calves that you are grazing on that particular farm and enter the data in the monthly column.

Right next to the monthly column I keep a yearly total of the calf grazing days. This is an interesting figure to compare one grazing year to the next. The total year calf grazing days is also a yardstick measure for a rough estimate of total weight gain as the grazing season progresses.

For example, if you want to be conservative on calculating how much a group of calves have gained, simply multiply 0.6 pounds x the total grazing days. (Example: 20,000 total calf grazing days x 0.6 lbs = 12000 pounds of gain). Take 12000 lbs x 0.32 cents (custom gain charge) = $3840. If you can get the calves to gain 1.2 lb per day you have doubled your paycheck, $7680.

This kind of math gets me excited and more determined than ever to manage my grass. By doing these calculations it really keeps me focused on how important it is to keep your daily gains as high as possible. Keep that grass legume sward vegetative and your bank account will prosper.

If you have multiple farms stocked with calves from one stocker owner, which I do at times, I have an additional spreadsheet that makes the monthly grazing record keeping simpler. This spreadsheet is titled "Total Calf Grazing Days Of All Farms." (See page 221.)

This sheet is where all the farms are listed with their present stocking number and calf grazing days for the month. All farm grazing days are totaled together for a grand total. This number is simply multiplied by 0.16 cents, which gives you the total amount due for the month.

One other spreadsheet I keep is called "Financial Review." This spreadsheet agenda covers all aspects of the farm: growth goals, profit, losses, future projects, accomplishments, liabilities and equity. Listed on the following pages is an example of what the "Financial Review" covers.

Financial Review 1999
Joe Smith's Custom Grazing Farms

Revenue:
1. Gross Farm Income 1999 = $45000
2. Net profit 1999 = $35000

Expenses:
1. Total For 1999 =$15000

Expense breakdown by category:
* Water establishment =
* Hay, mineral, salt =
* Utilities =
* Fencing materials =
* Lime, seed =
* Farm lease payment =
* Fuel =
* Misc. =

Revenue Breakdown By Farm:
1. Smith Farm =
2. Colley Farm =
3. Jones Farm =
4. Titus Farm =

Balance Sheet

List of Assets: (Example)
1. Cattle
2. Home
3. Land
4. Truck
5. ATV
6. Water tanks

7. Fences
8. Savings
9. Etc.

Liabilities (Debt)
1. Home loan
2. Cattle loan
3. Etc.

Current Ratio
= (Assets divided by liabilities)
Tip: The higher you can keep your current ratio, the better the banker will like you.

Quarterly Farm Update

1. Attained new grazing lease on Smith farm.
2. Paid down balance on livestock loan to $10,000.
3. Have completed a new land lease proposal on 100 acres that attaches to Smith farm.
4. Got contract signed on 150 steers for fall grazing.
5. Got all paddocks installed on Titus farm.
6. Have run above ground water pipe on Jones farm.
7. Last group of steers loaded off Smith farm averaged 1.35 lb per day.

New Growth Topics

1. Establish watering areas on Smith Farm
2. Select best performing heifers on grass to build a grass-finished direct-marketing beef product.
3. Start holding farm tours for local grade school kids to teach them about the environmentally friendly MiG practice.
4. Etc.

This financial review sheet helps keep me focused on all aspects of the grazing company. Revenue and expenses are recorded here. Following is the net profit. You need to be honest with yourself here. If your farms are not making a net profit, this review sheet will uncover why.

You also should break down the revenue by what is generated by each farm. This gives you a real accurate measure of which farms are performing the best. It also allows you to compare prior years to the present, which enables you to see the progress you're making.

The balance sheet is made up of your assets and liabilities. You also can calculate your "Current Ratio" from these two.

Example: Assume your total assets are $100,000. Assume your total liabilities are $20,000. Take $100,000 divided by $20,000 = current ratio of 5. The higher your current ratio is, the better off you are financially, especially in the eyes of the banker.

I do a quarterly update on the grazing company. I like this better because you are better informed about your grazing company by going over it quarterly rather than once at the end of the year. It gives you a chance to make adjustments if you find out something is not going the way that you had planned from an earlier quarterly update.

The last section, "New Growth Topics," is great for keeping an exciting list of new ideas flowing forward for implementation into the grazing company. The financial review sheet really helps keep me focused and informed as to where the grazing company has been, where it is at the present, and where it plans to be in the future. I highly recommend that you keep some form of financial review sheet for your own grazing company.

Chapter 26
Never Stop Learning!

As you can probably tell by now, I'm a huge advocate for always learning new ideas. A person should never be satisfied with thinking that they have learned everything there is to learn. By adopting an attitude that you know everything, you are doomed for growing any further in your endeavors.

First of all, once you think you know it all, your mind closes and refuses to receive new data. Secondly people start to think that you are a know it all and will not share any new ideas with you.

You must learn to be humble in life, be open minded enough to listen to other peoples' successes and failures. You can learn a lot just by listening to other people.

Any chance that I get to further my education in grazing, I do my best to take advantage of it. I have learned mountains of information from reading many of the good grazing books in many publications. If you are not reading the *Stock-*

man Grass Farmer every month, you are just plain missing a wealth of priceless grazing, marketing, cattle price cycles and state of the art forage information. You simply can not pick up any other monthly magazine anywhere in the world that gives you this kind of priceless data. No other publication can hold a candle to it.

Be open minded enough to listen to other people's successes and failures.

I have taken the magazine for about eight years. It simply keeps me on track as to where the cattle cycle is, whether I should be selling or buying, and keeps me informed on all the new grazing methods. It changes your thought process about the correct way to raise livestock. Heck, I've already told you that it was Allan Nation's article on leasing land that turned my whole life around. Just do it. Get yourself a subscription and I guarantee your grazing business will have a positive effect as a result of reading about the best graziers in the world every month. You simply owe it to yourself to be armed with all the forage knowledge that you can get your hands on.

Every grazing seminar that I get a chance to attend is a welcome treat. When I attend these seminars, I always bring an 8x12 inch notebook pad to take notes on from each speaker. When I get home these notes are filed under the seminar's name and put in my filing cabinet for future reference. I find myself constantly going back years later and going over my notes from these seminars. I have found that by taking notes during the speaker's presentation it implants the data in my mind better for later use. I also have a nice written record to refer back to later.

I know it is easier to just sit and listen intently to the speaker give their presentation. 90 percent of the people do just that. But go back a month later to those people who didn't take notes and see how much they retained from the presentation. If you're not much on writing, then just concentrate on jotting down the very key points.

Seek out experienced graziers who are performing grazing practices the way they should be done. A little advice when talking with these seasoned graziers, listen and take notes. Do not talk excessively about your own operation to the professional grazier. Be humble. Give him the greatest respect and he may tell you all the details of his success.

I have witnessed numerous inexperienced individuals at grazing seminars just get completely carried away with explaining a topic to one of the guest speakers. They will go into a ten minute story explaining every single little detail about a certain pasture or group of cows, etc. They absolutely hog the entire conversation and consequently they do not learn any key pointers from the professional grazier simply because they did not ask any questions.

It is very frustrating to be all primed up with a good juicy question to ask the guest speaker after their talk, yet never get a chance because someone is telling the speaker all about their farm. Remember you can learn a lot more by listening than you can by talking. Your brain is more receptive if you are still, yet tuned in to what is being said.

Once you have some success with leasing and grazing, do not let it go to your head. I've met a lot of people who looked down their nose at me after they became successful. What a travesty, because they are still the same person they were before they were successful, but now they act like they don't know your name. By assuming that you are this almighty high and important person in society

Seek experienced graziers who are performing grazing practices the way they should be done.

now, you are putting a lot of needless stress on yourself. Let me explain further in detail. Once you assume you are an important person, then every time someone does not give you the respect that you think you deserve, it upsets you. So you are constantly trying to defend this worthless important person title that you have placed on yourself. Do not waste good energy playing the important person role. Learn to be humble. Laugh at your mistakes and enjoy life.

Give the seasoned MiG grazier the respect he deserves and he may tell you all the details of his success.

When it came to teaching, I had an uncle who was the best. His name was Scott Judy, a bachelor who traveled around the world twice while serving in the Peace Corps. He taught in Africa for several years. He lived among the villagers and taught them how to raise crops, livestock, marketing, water establishment and taught their children some educational skills.

He owned a 700 acre grass farm that he worked at like a soldier every time he got the chance. He was always brush hogging and building fence every time we would go see him. He taught me how to tamp post holes when I was seven years old. He was very diligent about getting every morsel of dirt back in that post hole. I told him that there was no way that all that dirt could go back in that hole with that big old post taking up all that space. He patiently explained why the dirt would fit if we tamped it correctly.

Every time he came by our house it was always a treat because he would always teach us something new. He would teach numerous children how to read and write before they ever started school. He was always quizing you, trying to get you to use your brain power. Uncle Scott never acted important or talked down to you. He had a nack of bringing the best out in everybody. He never talked bad about anybody and was always willing to lend a helping hand to an absolute stranger if he thought he needed it.

One of my old neighbors never saw a sober day in 37 years until my Uncle Scott got him to go into a hospital to dry out. Shortly after the old neighbor got out of the hospital, Uncle Scott gave him a house on one of his farms to live in as long as he promised to stay sober. Uncle Scott was killed shortly after this by a drunken driver. The old ex alcoholic never touched a drop of liquor the rest of his life. Lots of people had tried to get the old man to dry out, but all were unsuccessful until Scott came along.

The old alcoholic told me later that Scott made him feel so important and put so much faith in him that he was bound and determined not to let Scott down, and by josh he didn't. The old man told me that Uncle Scott was the first man that he ever met who wanted to do something good for him and never asked for anything in return.

Today I own 200 acres of my Uncle Scott's farm. I really believe that if my Uncle Scott was alive today he would be very proud of the way his farm looks now and the way that it is being managed. I learned a lot from Uncle Scott in the short time that I got to be around him.

The one thing that he burned into my soul was to always keep learning new information and topics in this vast exciting constantly changing world that we live in.

Grazier's Glossary

Animal unit day: Amount of forage necessary to graze one animal unit (one dry 1100 lbs beef cow) for one day.

Annual leys: Temporary pastures of annual forage crops such as annual ryegrass, oats or sorghum-sudangrass.

AU Animal unit: One mature, non-lactating cow weighing 500kgs (1100 lbs) or its weight and class equivalent in other species. (Example: 10 dry ewes equal one animal unit.)

AUM: Animal unit month. Amount of forage needed to graze one animal unit for a month.

Blaze graze: A very fast rotation used in the spring to prevent the grass from forming a seedhead. Usually used with dairy cattle.

Break grazing: The apportioning of a small piece of a larger paddock with temporary fence for rationing or utilization purposes.

Breaks: An apportionment of a paddock with temporary electric fence. The moving of the forward wire would create a "fresh break" of grass for the animals.

Carrying capacity: Stocking rate at which animal performance goals can be achieved while maintaining the integrity of the resource base.

Composting: The mixing of animal manure with a carbon source under a damp, aerobic environment so as to stabilize and enhance the nutrients in the manure.

Continuous stocking: Allowing the animals access to an entire pasture for a long period without paddock rotation.

Compensatory Gain: The rapid weight gain experienced by animals when allowed access to plentiful high quality forage after a period of rationed feed. Animals that are wintered at low rates of gain and are allowed to compensate in the spring frequently weigh almost the same by mid-summer as those managed through the winter at a high rate of gain. Compensatory gain is sometimes also known as "pop."

Creep grazing: The allowing of calves to graze ahead of their mothers by keeping the forward paddock wire high enough for the calves to go under but low enough to restrain the cows.

Name ____
Address ____
City ____
State/Province__________________Zip/Postal Code _______
Phone __________________________

Quantity	Title	Price Each	Sub Total
____	**20 Questions** (weight 1 lb)	**$22.00**	______
____	**Comeback Farms** (weight 1 lb)	**$29.00**	______
____	**Creating a Family Business** (weight 1 lb)	**$35.00**	______
____	**Drought (weight 1/2 lb)**	**$18.00**	______
____	**Grassfed to Finish** (weight 1 lb)	**$33.00**	______
____	**Kick the Hay Habit** (weight 1 lb)	**$27.00**	______
____	**Kick the Hay Habit Audio - 6 CDs**	**$43.00**	______
____	**Knowledge Rich Ranching** (wt 1½ lb)	**$32.00**	______
____	**Land, Livestock & Life** (weight 1 lb)	**$25.00**	______
____	**Management-intensive Grazing** (wt 1 lb)	**$31.00**	______
____	**Marketing Grassfed Products Profitably** (1½)	**$28.50**	______
____	**No Risk Ranching** (weight 1 lb)	**$28.00**	______
____	**Paddock Shift** (weight 1 lb)	**$20.00**	______
____	**Pa$ture Profit$ with Stocker Cattle** (1 lb)	**$24.95**	______
____	**Pa$ture Profit abridged Audio -- 6 CDs**	**$40.00**	______
____	**Quality Pasture** (weight 1 lb)	**$30.00**	______
____	**The Calendar of Year-round Grazing**	**$18.00**	______
____	**The Moving Feast** (weight 1 lb)	**$20.00**	______
____	**The Use of Stored Forages (weight 1/2 lb)**	**$18.00**	______
____	Free Sample Copy ***Stockman Grass Farmer*** magazine		______
		Sub Total	______

Mississippi residents add 7% Sales Tax _______ Postage & handling _______

Shipping	Amount
1/2 lb	$3.00
1- 2 lbs	$5.60
2-3 lbs	$7.00
3-4 lb s	$8.00
4-5 lbs	$9.60
5-6 lbs	$11.50
6-8 lbs	$15.25

Canada
1 book $18.00
2 books $25.00
3 to 4 books $30.00

TOTAL _______

Foreign Postage:
Add 40% of order

We ship 4 lbs per package maximum outside USA.

www.stockmangrassfarmer.com

Please make checks payable to

Stockman Grass Farmer
PO Box 2300
Ridgeland, MS 39158-2300

1-800-748-9808
or 601-853-1861
FAX 601-853-8087

More from Green Park Press

MARKETING GRASSFED PRODUCTS PROFITABLY by Carolyn Nation. From farmers' markets to farm stores and beyond, how to market grassfed meats and milk products successfully. Covers pricing, marketing plans, buyers' clubs, tips for working with men and women customers, and how to capitalize on public relations without investing in advertising. 368 pages. **$28.50***

NO RISK RANCHING, Custom Grazing on Leased Land by Greg Judy. Based on first-hand experience, Judy explains how by custom grazing on leased land he was able to pay for his entire farm and home loan within three years. 240 pages. **$28.00***

PADDOCK SHIFT, Revised Edition Drawn from Al's Obs, Changing Views on Grassland Farming by Allan Nation. A collection of timeless Al's Obs. 176 pages. **$20.00***

PA$TURE PROFIT$ WITH STOCKER CATTLE by Allan Nation. Profiles Gordon Hazard, who accumulated and stocked a 3000-acre grass farm solely from retained stocker profits and no bank leverage. Nation backs his economic theories with real life budgets, including one showing investors how to double their money in a year by investing in stocker cattle. 192 pages **$24.95*** or Abridged audio 6 CDs. **$40.00**

QUALITY PASTURE, How to create it, manage it, and profit from it, Revised 2nd Edition, by Allan Nation and Jim Gerrish. Down-to-earth, low-cost tactics to create high-energy pasture that will reduce or eliminate expensive inputs or purchased feeds. 300 pages **$30.00***

THE CALENDAR OF YEAR-ROUND GRAZIER by Steve Kenyon. Although Kenyon only has a 4 month grazing season, he explains how to graze year-round in any region. Includes economic and financial information to help make your operation profitable 12 months of the year. 140 pages **$18.00***

THE MOVING FEAST, A cultural history of the heritage foods of Southeast Mississippi by Allan Nation. How using the organic techniques from 150 years ago for food crops, trees and livestock can be produced in the South today. 140 pages. **$20.00***

THE USE OF STORED FORAGES WITH STOCKER AND GRASS-FINISHED CATTLE. by Anibal Pordomingo. Helps determine when and how to feed stored forages. 58 pages. **$18.00***

* All books softcover. Prices do not include shipping & handling

To order call 1-800-748-9808

or visit www.stockmangrassfarmer.com

Index

Standing hay: The deferment of seasonally excess grass for later use. Standing hay is traditionally dead grass. Living hay is the same technique but with green, growing grass.

Stock density: The number of animals on a given unit of land at any one time. This is traditionally a short-term measurement. This is very different from stocking rate which is a long term measurement of the whole pasture. For example: 200 steers may have a long-term stocking rate of 200 acres, but may for a half a day all be grazed on a single acre. This acre while being grazed would be said to have a stock density of 200 steers to the acre.

Stocker cattle: Animals being grown on pasture between weaning and final finish. Stocker cattle weights are traditionally from 350 to 850 lbs.

Stocking rate: A measurement of the long-term carrying capacity of a pasture. See stock density.

Stockpiling: The deferment of pasture for use at a later time. Traditionally this is in the autumn. Also known as "autumn saved pasture" or "foggage."

Stripgraze: The use of a frequently moved temporary fence to subdivide a paddock into very small breaks. Most often used to ration grass during winter or droughts.

Swathed oats: The cutting and swathing of oats into large double-size windrows. These windrows are then rationed out to animals during the winter with temporary electric fence. This method of winter feeding is most-often used in cold, dry winter climates.

Value of Gain: The net value of gain after the price rollback of light to heavy cattle has been deducted. To find the net value of gain, the total price of the purchased animal is subtracted from the total price of the sold animal. This price is then divided by the number of cwts. of gain. Profitability is governed by the value of gain rather than the selling price per pound of the cattle.

VDMI: Voluntary dry matter intake. What the animal will consume on its own.

Wintergraze: Grazing in the winter season. This can be on autumn saved pasture or on specially planted winter annuals such as cereal rye and annual ryegrass.

Management-intensive Grazing (MiG): The thoughtful use of grazing manipulation to produce a desired agronomic and/or animal result. This may include both rotational and continuous stocking depending upon the season.

Mixed grazing: The use of different species grazing either together or in a sequence.

Mob grazing: A mob is a group of animals. This term is used to indicate a high stock density.

N: Nitrogen

P: Phosphorus.

Paddock: A permanently fenced subdivision of a pasture.

Pastureland: Land used primarily for grazing purposes.

Pugging: Also called bogging. The breaking of the sod's surface by the animals' hooves in wet weather. Can be used as a tool for planting new seeds.

Residue: Forage that remains on the land after a harvest.

Residual: The desired amount of grass to be left in a paddock after grazing. Generally, the higher the grass residue, the higher the animal's rate of gain and milk production. This can be expressed as either sward height or forage mass. Example: We could have a four-inch residual or a 1500lb/acre residual.

Ruminants: Hooved livestock with multiple stomachs and which chew their cud.

Seasonal grazing: Grazing restricted to one season of the year. For example, the use of high mountain pastures in the summer.

Self feeding: Allowing the animals to eat directly from the silage face by means of a rationing electric wire or sliding headgate.

Set stocking: The same as continuous stocking. Small groups of animals are placed in each paddock and not rotated. Frequently used in the spring with beef and sheep to keep rapidly growing pastures under control.

Split-turn: The grazing of two separate groups of animals during one grazing season rather than one. For example, the selling of one set of winter and spring grazed heavy stocker cattle in the early summer and the replacement of them with lighter cattle for the summer and fall.

Spring flush or lush: The period of very rapid growth of cool-season grasses in the spring.

CWT: 100 pounds.

Deferred grazing: The dropping of a paddock from a rotation for use at a later time.

Dirty Fescue: Fescue containing an endophyte which lowers the animal's ability to deal with heat. Fescue without this endophyte is called Fungus-free or Endophyte-free.

Dry matter: Forage after the moisture has been removed.

Flogging: The grazing of a paddock to a very low residual. This is frequently done in the winter to stimulate clover growth the following spring.

Forbs: General term used to describe broad-leafed plants.

Frontal Grazing: An Argentine grazing method whereby the animals' grazing speed is determined with the use of a grazing speed governor on a sliding fence.

Grazer: An animal that gathers its food by grazing.

Grazier: A human who manages grazing animals.

Grazing pressure: How deep into the plant canopy the animals will graze.

Intake: Amount of forage an animal will consume.

K: Potassium.

Lax grazing: The allowing of the animal to have a high degree of selectivity in their grazing. Lax grazing is used when a very high level of animal performance is desired.

Leader/follower: A leader/follower grazing system is one in which two or more classes of livestock having distinctly different nutritional needs or grazing habits are grazed successively in a pasture.

Legumes: Plants that bear fruits such as beans or clover.

Ley pasture: Temporary pasture. Usually of annuals.

Leader/follower Grazing: The use of a high production class of animal followed by a lower production class. For example, lactating dairy cattle followed by replacements. This type of grazing allows both a high level of animal performance and a high level of pasture utilization. Also, called first-last grazing.

Lodged over: Grass that has grown so tall it has fallen over on itself. Most grasses will self-smother when lodged. A major exception is Tall fescue and for this reason it is a prized grass for autumn stockpiling.